Materials Science and Technologies

Challenges and Opportunities in the Textile Industry

MATERIALS SCIENCE AND TECHNOLOGIES

Additional books and e-books in this series can be found on Nova's website under the Series tab.

MATERIALS SCIENCE AND TECHNOLOGIES

CHALLENGES AND OPPORTUNITIES IN THE TEXTILE INDUSTRY

WALLACE G. TARRANT
EDITOR

Library of Congress Cataloging-in-Publication Data

ISBN: 978-1-53618-770-0

Published by Nova Science Publishers, Inc. † New York

CONTENTS

PREFACE

This compilation presents a detailed review of current research, developments, and progress on nanotechnology usage for the elimination of dyes from effluents released by textile industries.

The benefits of using nanomaterials for functionalized textile production are presented, and the applications of nanomaterials in the most known functional technical textiles are discussed.

The authors present the results of empirical studies carried out in the Portuguese industrial context, including the textile sector, where the relationship between negative effects associated with shift work and the adoption of certain management practices by organizations is analyzed.

Additionally, the authors discuss how, to achieve the Fourth Industrial Revolution, technological tools must be incorporated into both the production and consumption of textiles.

The closing study indicates that it is possible to make objective pilling detection easily for standard fabric structures in the textile industry using databases created by measuring lots of samples.

Chapter 1 - World textile production has been reliably expanding in recent years. Treatment of wastewater is one of the crucial tasks in the textile industry. Improper waste management may lead to environmental problems as well as a health hazard to aquatic life, which may cause an imbalance in Bio-system. Hence, the reduction of waste generation or elimination of waste from the effluents of textile industries is essential worldwide. Some of the efficient ways for waste removal from effluents (mainly water) are Nano-filtration, electrochemical oxidation, electrocoagulation–electro-oxidation, etc. In any textile industry, along with making the use of various metals and other components, the utilization of dyes is considered as an integrated part of it, to color the fabrics. These dyes, metals, and other pollutants need to be removed from the industrial wastewater and therefore has gained large attention in recent years due to the enforcement of strict environment laws, which have impelled the industries and factories to treat their effluents before releasing them into freshwater resources. Various nanotechnology-based materials, such as nano-membranes, nano-metals, and nano-adsorbents are successfully used for removing pollutants from the wastewater. With the rapid progress of nanotechnology, there is no doubt that in the next few years nanotechnology will penetrate every area of the textile industry. The current chapter presents a detailed review of current researches, developments, and progresses on nanotechnology usage for the elimination of dyes from the effluents released by textile industries. Usage and performances of different nanostructured materials such as nanoparticles, nanocomposite, nanofiltration, nanofibrous membrane on dye removal are also reviewed and discussed comprehensively.

Chapter 2 - Nanosized materials are becoming more and more popular in almost every field of the industry. Technical textiles are one of the major applications of nanomaterials and there are many commercialized nanoparticles functionalized textiles on the market such as electromagnetic interference shielding materials, flame retardant textiles, UV-protective fabrics, etc. The compatibility of conventional textile processes for nanoparticle applications and reduced nanomaterial production costs are made much faster to produce commercialized nanoparticle functionalized textiles. Nanoclays, carbon-based nanomaterials, and metal-based nanopowders are the most commonly used nanomaterials in functional textiles. It is possible to produce nanomaterial functionalized textile by adding nanoparticles into polymer melt or solution for synthetic fibres, or surface treatment of fiber or textile materials. In this chapter of the book, the benefits of using nanomaterials for functionalized textile production will be presented. Applications of nanomaterials in the most known functional technical textiles, (flame retardant, antimicrobial, electromagnetic interference shielding, UV-protective, superhydrophobic, and wrinkle resistive) will be discussed.

Chapter 3 - The use of shift work is a practice adopted by many organizations, covering a significant number of workers. For example, according to Eurostat (2020), in 2019, 18.4% of workers in the 28 Member States of the European Union worked shifts; in Portugal, the value is slightly higher: 19.7%. On the other hand, there are significant differences in the use of this work schedule between sectors of activity. This is prevalent, for example, in the field of health, commerce and hospitality or in some industrial sectors (Eurofound, 2016).

In general, shift work, especially that involving night shifts, has been associated with several negative consequences for workers and organizations. This chapter is dedicated to reviewing the main consequences associated with shift work for shift workers (health and family and social life) and for organizations (safety and performance). This review will focus on studies carried out in an industrial context in general, and in the textile sector in particular. The chapter ends with recommendations on the management of shift work schedules, where the importance of flexible practices in this management will be emphasized. In this topic, it will also be presented the results of empirical studies carried out under the scientific coordination of the first author, in the Portuguese industrial context, including the textile sector, where the relationship between negative effects associated with shift work (e.g., health) and the adoption of certain management practices by organizations is analyzed (e.g., worker participation in the management of working hours).

Chapter 4 - Each Industrial Revolution played a key role in the production-consumption pattern of textile products. Given the complexity of the Textile and Apparel Industry, the production systems have varying degrees of conceptual and technological innovation. Thus, the objective of this research is to present a tool that integrates Life-cycle Assessment and Industry 4.0. For this, the analysis was delimited to four production systems for the manufacturing of 100% combed cotton t-shirts: obtaining of textile fibers, spinning, knitting and garment confection. The LCA indicated a participation of all systems analyzed in the generation of negative impacts on human health, climatic, atmospheric and soil conditions, but the stage of transformation of fibers into wires presented the worst environmental indexes. From the technological monitoring, it was possible to infer that the conceptual novelty and complexity of configuration of the components are in an evolutionary process, in which there is automation of machine parts and processes. To achieve the Fourth Industrial Revolution, technological tools must be incorporated into both the production and consumption of textiles.

Chapter 5 - Pilling is a serious defect of fabric surface that gives an unpleasant appearance to garment. Pilling tendency is tested with different methods and devices in the laboratory conditions. The determination of the pilling grades is made with visual control by operators. Therefore, the human factor is significantly effective in this subjective evaluation method and may cause incorrect results. Studies in recent years show that objective methods based on image processing are preparing to replace subjective pilling assessments. In this chapter, difficulties in the subjective evaluation of the pilling grades were explained. Potential opportunities presented by image processing studies in the literature on the objective evaluation of the pilling grades were investigated. Image processing steps were given with various examples by using Image Processing Toolbox and codes in MATLAB software. In this study, it was indicated that it is possible to make an objective pilling detection easily for the fabric structures used as standard in the textile industry thanks to the databases to be created with measuring lots of samples.

In: Challenges and Opportunities … ISBN: 978-1-53618-770-0
Editor: Wallace G. Tarrant

Chapter 1

NANOTECHNOLOGY FOR DYE REMOVAL FROM TEXTILE INDUSTRY EFFLUENTS

Manash Protim Mudoi*[1]*, Shilpi Agarwal*[2,*]*,
***Shailey Singhal*[2]*, Abhimanyu Singh Khichi*[1]**
***and S. Dayanidy*[1]**
[1]Department of Chemical Engineering,
University of Petroleum and Energy Studies, Dehradun, India
[2]Department of Chemistry,
University of Petroleum and Energy Studies, Dehradun, India

ABSTRACT

World textile production has been reliably expanding in recent years. Treatment of wastewater is one of the crucial tasks in the textile industry. Improper waste management may lead to environmental problems as well as a health hazard to aquatic life, which may cause an imbalance in Bio-system. Hence, the reduction of waste generation or elimination of waste from the effluents of textile industries is essential worldwide. Some of the efficient ways for waste removal from effluents (mainly water) are Nano-

* Corresponding Author's Email: shilpi@ddn.upes.ac.in.

filtration, electrochemical oxidation, electrocoagulation–electro-oxidation, etc. In any textile industry, along with making the use of various metals and other components, the utilization of dyes is considered as an integrated part of it, to color the fabrics. These dyes, metals, and other pollutants need to be removed from the industrial wastewater and therefore has gained large attention in recent years due to the enforcement of strict environment laws, which have impelled the industries and factories to treat their effluents before releasing them into freshwater resources. Various nanotechnology-based materials, such as nano-membranes, nano-metals, and nano-adsorbents are successfully used for removing pollutants from the wastewater. With the rapid progress of nanotechnology, there is no doubt that in the next few years nanotechnology will penetrate every area of the textile industry. The current chapter presents a detailed review of current researches, developments, and progresses on nanotechnology usage for the elimination of dyes from the effluents released by textile industries. Usage and performances of different nanostructured materials such as nanoparticles, nanocomposite, nanofiltration, nanofibrous membrane on dye removal are also reviewed and discussed comprehensively.

Keywords: wastewater, dyes, nanotechnology

1. Introduction

Now a day's textile industry has become one of the major and most complex industrial chains in the manufacturing sector. Considerable amounts of effluents are generated which can cause serious environmental problems if disposed off without proper treatment. Among different wastes, dyes are the most released pollutants. About tens of thousands of various pigments and dyes are commercially used in the textile industry. Discharge of dyes into water bodies poses mutagenic and carcinogenic effects to human and aquatic life because of the presence of aromatic rings in most of the dyes. Therefore, removing dyes from contaminated water from textile industries is very crucial to prevent its harmful effects, and therefore, the focus on the development of efficient dye removal techniques is increasing considerably (Drumond Chequer et al. 2013). Traditional processes for water & wastewater treatment are becoming ineffective due to the increase in freshwater demand and stringent guidelines for freshwater. In recent years, nanotechnology-based processes have attracted huge attention in the treatment of wastewater because of its high efficiency in removing dyes and its technological advances. Numerous researchers reported the development of various types of nanoparticles, nanofibers, nanocomposites, nanofibrous membranes which can efficiently process many of the water contaminants from the textile industry (Savage and Diallo 2005). In this chapter, we have summarized the role of various nanomaterials in wastewater treatment to conquer the water crisis.

2. Textile Dyes and Its Impact

Dyes are chemical compounds, which act as a coloring agent by binding chemically to the material. They are organic material consisting of chromophores and auxochromes, which are responsible for their color. Chromophores impart the color to the dye while auxochromes enhance the intensity of dye color. Among many typcs, nitroso (-NO) azo (-N=N-), nitro (-NO_2), carbonyl (C=O) are some major chromophores and the hydroxyl (-OH), carbonyl (C=O), amino (-NR_2), amine (-NH_2), sulfonate (-SO_3H) groups are the most important auxochromes (Allen 1971). Generally, dyes are soluble in water and give light color in the water, however, some dyes produce very dark colors even at very low concentrations. Dyes can either be synthetic or natural. Production of natural dyes is mainly done from the extract of vegetables, fruits, flowers, insects, and natural minerals, etc. Their drawbacks are their limited range of color choices as well as ecosystem imbalance since the process disturbs the natural pollination and reproduction cycle of plant species. Also, the natural dyes give a muted range of colors which fades when exposed to sunlight and washing. To overcome this problem, mordants are used widely which helps the dyes to fix firmly on the fabric. Contrary to the natural dyes, synthetic dyes are highly stable to heat and light. The first synthetic dye was “Mauveine,” which was discovered in 1856 by William H. Perkin, by oxidation of an aniline bases mixture. Thereafter, its success inspired the chemists to conduct extensive experiments to create synthetic dyes. In this way, a wide range of synthetic dyes with different brilliant, colorfast tones for various uses was invented. According to the Color Index (C.I), currently, more than tens of thousands of different types of dyes have produced. Worldwide the annual dye production is reported to be more than seven hundred thousand tonnes. Due to their complex and stable structure, synthetic dyes do not degrade upon contact with water, detergents, or any other washing agents and are used widely in all the textile industries however, synthetic dye components are mutagenic, toxic and carcinogenic in nature. They increase the COD (chemical oxygen demand) level and BOD (biochemical oxygen demand) level in aquatic sources threatening to the aquatic as well as human life. Therefore, it becomes very essential to remove synthetic dyes properly before discharging into the environment.

3. Classification of Dyes

Classification of dyes can be done mainly in two different ways: natural and synthetic (man-made dyes). Minerals, plants, and animals are the main sources of natural dyes whereas petrochemicals are the main ingredients for preparing the synthetic dyes. The finest way of classification of dyes is by its chemical structure because it readily recognizes them belonging to a particular class having distinctive properties. Also, it is the most widely used classification by synthetic dye chemists and dye experts. Classification of dyes based on their chemical structure and chromophores present is shown in Table 1. Sometimes it is also classified based on its solubility in water, like mordant, acidic, basic, metal complex, reactive, and direct dyes are water-soluble whereas vat, disperse, and sulfur dyes are water-insoluble (Hunger 2002).

Table 1. Dye classification based on the structure and chromophores present

Dyes	Chromophores	Example
Indigo	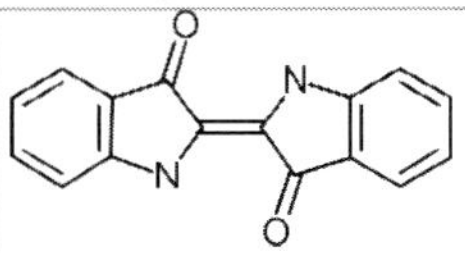	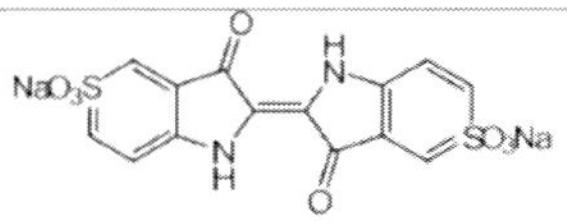 Acid blue 71
Anthraquinone	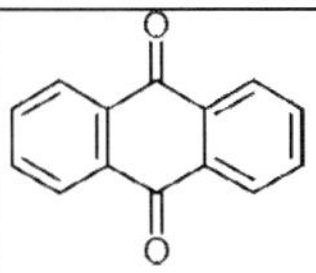	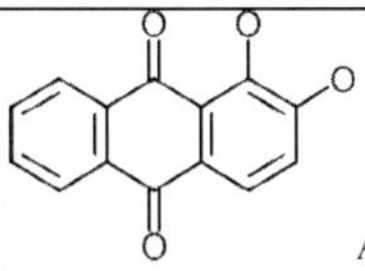 Alizarin
Triarylmethane	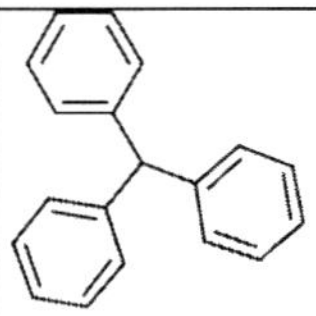	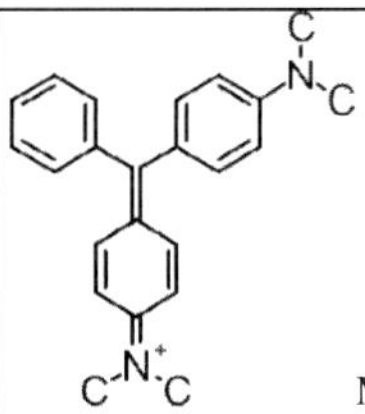 Malachite green
Azo	-N=N-	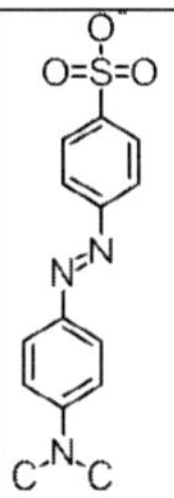 Methyl orange
Nitroso	R-N=O	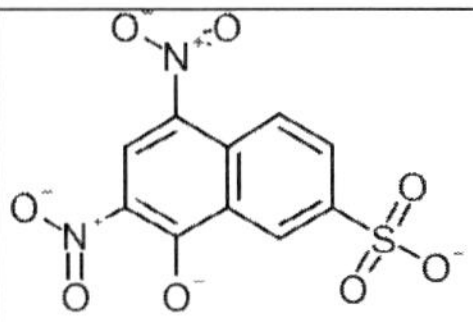 Naphthol yellow s
Nitro	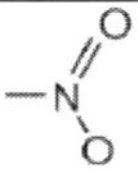	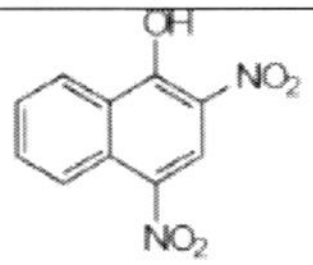 Acid Yellow 24

4. Nanotechnology for Dye Removal

From the past few decades, various chemical, biological, and physical methods are developed for decolorization of textile dye effluents. These include adsorption, coagulation-flocculation, solvent extraction, ion-exchange method, oxidation, ozonation, catalytic degradation, ultrafiltration, microfiltration, electrochemical destruction, electro-coagulation, aerobic or anaerobic treatment, and microbial degradation (Ejder-Korucu et al. 2015). However, these methods have their drawbacks. High electrical energy and chemical requirements, low efficiency in removing dyes, and a large volume of sludge production restrict the use of most of these methods. Among all the above methods, the most suitable method is adsorption because of its ease of operation and availability of various adsorbents. Various adsorbents such as polymeric resins, zeolites, bio-sorbents, activated carbons, activated alumina or silica gel, agricultural wastes, etc. have been used widely for wastewater treatment. Almost all the dye pollutants present in water can be removed using the above mentioned adsorbents but are not much effective in removing pollutants present in very low concentration. High cost and low reusability of these adsorbents also limit their use in wastewater treatment (Crini et al. 2019). Therefore, there is an urgent need to improve most of the traditional methods to obtain efficiency, faster rate, and economical operation. Nanotechnology has emerged as one of the leading and most promising technologies with a huge capability for treating textile dye effluents more efficiently than the traditional dye removal methods because of the characteristic physical properties of nanoscale material. Nanomaterials are those materials where at least one dimension is smaller than 100 nm (Qu, Alvarez, and Li 2013). Nanomaterials can possess remarkable properties different from their larger counterparts due to their very small size. Nanomaterials typically have size-dependent continuous properties like high reactivity, degree of functionalization, large surface area, high sorption capacity, fast dissolution, and discontinuous properties like superparamagnetism, quantum confinement, localized surface plasmon resonance, etc., which makes them suitable for applications in textile wastewater treatment (Khajeh, Laurent, and Dastafkan 2013). Structural modification of nanoparticles with the advanced tools has found its usage in different industrial and environmental applications. Because of its nano-scale size, nanoparticles can go into deeper and attack the contaminants, which is generally difficult to be done by conventional technologies (Amin, Alazba, and Manzoor 2014). A variety of nanomaterials such as nano adsorbents, nanomembranes, nanocatalysts, nanotubes, nanocomposites, etc. have been extensively used in the purification of textile wastewater by removing the different types of dyes. Table 2 summarizes some of the materials developed at the nanoscale to use in dye removal processes.

Table 2. Nanotechnology-based material for dye removal

Nano Material	Type	Usage	Reference
Mixed oxide nanoparticle	Adsorbent	Methylene Blue (MB) and Rhodamine 6G removal	(Pal et al. 2016; Agboola et al. 2018)
Nanocomposites	TiO_2-adsorbent nanocomposites	Methylene Blue removal	(Zhang, Zou, and Wang 2010)
Nanofiltration	Membrane-based	Reactive Black 5 (RB5) removal	(Cao et al. 2020)
Nanocatalysts	photocatalysts, Fenton-based catalysts, electrocatalysts	Azo dye removal	(X. Zhao et al. 2011; Agboola et al. 2018)
Nanofibrous membrane	Membrane-based, adsorbent	C.I. Basic Violet 14 removal	(El-Aassar et al. 2016)
Nano alloys	Adsorbent	All type of dye removal	(Girgis and Aziz 2013; Agboola et al. 2018)
Carbon nanotubes	Adsorbent	Removal of extremely small amount dyes (in ppb) and heavy metals	(Sadegh et al. 2017; Agboola et al. 2018)
Bio adsorbent (with chitosan-dendrimer nanostructure)	Adsorbent	Reactive Red 198 (RR198) and Reactive Black 5 (RB5) removal	(Sadeghi-Kiakhani, Arami, and Gharanjig 2013; Agboola et al. 2018)

5. NANOPARTICLES

Among various treatment processes like membrane treatment, ion exchange coagulation, oxidation, and electrochemical processes, adsorption is emerged as an effective dye removal technique because of the biological and chemical stability of most industrial dyes. In recent years, studies have conducted to develop adsorbents that can exhibit high adsorption capacity and low operational cost. In this regard nanoparticles with unique properties have received much attention (Monsef Khoshhesab and Souhani 2018).

Batch experiments are done to examine the stability of each nanomaterial on target dye-pollutant by studying the effect of various parameters such as adsorbent dosage, initial dye concentration, and pH of dye solution. After performing the experiments, the thermodynamic, kinetic, and isotherm study of a particular adsorption process can be done using the experimental data (Kim and Kim 2018).

The adsorbate (dye) quantity adsorbed by the adsorbent at a constant temperature with varying pressure can easily be presented by a curve known as adsorption isotherm. Isotherm models such as the Freundlich model and Langmuir model are the frequently used isotherm equations for studying the adsorption mechanism. Assuming surface heterogeneity, monolayer adsorption occurs in the Langmuir isotherm model whereas multilayer adsorption can be considered for the Freundlich isotherm model (de Sá et al. 2017). The rate of adsorption is determined by adsorption kinetics study and it also foretells the adsorbent-adsorbate interaction depending on solute concentration, flow, and surface complexity of the adsorbent. The kinetics of the adsorption process is analyzed with pseudo-first-order and pseudo-second-order models. The thermodynamic parameters such as the entropy change, the enthalpy change, and the Gibbs' free energy change are calculated using the Van't Hoff's equation. The change in Gibb's free energy foretells the spontaneous nature of a process. Table 3 shows some nanoparticles used in the removal of dye along with their adsorption capacity, isotherm, and kinetics.

Table 3. Nanoparticles as an adsorbent for dye removal

Sl. no	Adsorbent	Dye	Adsorption capacity(mg/g)	Model*	Kinetics**	Reference
1.	Bimetallic (Agar-based) nanoparticles	Methylene Blue	0.9285	F	PFO	(Patra et al. 2016)
2.	MgO nanoparticle	Indanthrene Blue	86.50	L	PSO	(Venkatesha et al. 2012)
3.	$FeMgO_{IM}$	Remazol Red 133	36.9	L	PFO	(Mahmoud, El-Molla, and Saif 2013)
4.	$FeMgO_{Co}$	Remazol Red 133	23.1	L	PFO	
5.	$FeMgO_{HY}$	Remazol Red 133	32.3	L	PFO	
6.	MgO nanoparticle	Leva fix fast red CA	92.16	L	PSO	(Venkatesha et al. 2012)
7.	ilmenite $FeTiO_3$	Methylene Blue	71.9	L	PFO	(Y. H. Chen 2011)
8.	Nano-Fe_3O_4	Methylene Blue	105	L	PSO	(Iram et al. 2010)
9.	nano-MgO	Reactive Blue 19	166.7	L	PSO	(Moussavi and Mahmoudi 2009)
10.	nano-MgO	Reactive Red 198	123.5	L	PSO	
11.	Co_3O_4/SiO_2	Methylene Blue	53.87	L	FO	(Kannan and Sundaram 2001)
12.	Nano-ZnO	Reactive Blue 19 and Acid Black 210	38.02 and 34.13	L	PSO	(Monsef Khoshhesab and Souhani 2018)
13.	Nano-ZnO	Azo dyes	-	L	PSO	(Zafar et al. 2019)
14.	Fe_3O_4@L-arginine	Reactive Blue 19 (RB19)	-	F	PSO	(Dalvand et al. 2016)
15.	Nano-NiO	Methyl Orange	97.56	L	PSO	(Riaz et al. 2020)

* L: Langmuir isothermal model.
F: Freundlich isothermal model.
S: Surfactant.
** PFO: pseudo-first-order.
PSO: pseudo-second-order.
FO: first-order.
Fe_3O_4@L-arginine: Fe_3O_4 (iron oxide) magnetic nanoparticles modified with L-arginine.

Patra et al., 2015 synthesized some agar-based monometallic (Pd, Cu, and Fe) and bimetallic (Fe/Pd and Fe/Cu) nanoparticles for removing the

Methylene Blue (MB) and Rhodamine B dyes. After performing experiments and analyzing the results, it was observed that for agar-based Fe/Pd nanoparticle, the maximum adsorption uptake was 875.0 mg/g for Rhodamine B and 780.0 mg/g for MB. The adsorption data were best explained with the pseudo-first-order kinetic model and Freundlich isotherm model. Also, the adsorbent was renewable and used for almost 20 cycles to get degradation rate of 90% and more (Patra et al. 2016).

Dalvand et al., 2016 utilized Fe_3O_4 magnetic nanoparticles modified with L-arginine as an adsorbent for removing Reactive Blue 19 (RB19) dye from textile dye effluents. A degradation rate of 96.34% was achieved with a 50 mg/l concentration of dye and a 0.74 g/l concentration of adsorbent at pH 3. The Freundlich isotherm was found to be the most appropriate model and the adsorption kinetics were explained with the pseudo-second-order model (Dalvand et al. 2016).

Monsef Khoshhesab and Souhani et al., 2018 synthesized ZnO particles using the precipitation method. They have discussed the physical properties and dye adsorption characteristics of ZnO nanoparticles and applied for removal of RB19 dye (anthraquinone dye) as well as Acid Black 210 (AB210) dye. The highest adsorption uptake reached 38.02 mg/g and 34.13 mg/g for RB19 and AB210 respectively, as determined from Langmuir isotherm. Pseudo-second-order kinetic model was the best chosen model for kinetic analysis (Monsef Khoshhesab and Souhani 2018).

Zafar et al., 2019 prepared ZnO nanoparticles using the co-precipitation method and used for removing azo dyes. After doing the experiments, the results revealed the maximum removal efficiency of 40 ppm for every 0.3 g of nanoparticles used. Langmuir isotherm equation and the pseudo-second-order kinetic model were suitable for fitting the adsorption data without much deviation (Zafar et al. 2019).

Riaz et al., 2020 synthesized NiO nanoparticles using a green hydrothermal method and utilized as an adsorbent for removing Methyl Orange (MO) dye. The maximum adsorption capacity for MO was 97.56 mg/g for 0.03 g of adsorbent used at a pH of 4. Adsorption of MO dye on NiO nanoparticles follows the Langmuir isotherm model and pseudo-second-order model (Riaz et al. 2020).

6. NANOCOMPOSITES

Carbon nanotubes (CNTs) find extensive use for removing different dyes because of their strong affinity for the dye molecules. It has an additional advantage of removing a toxic dye selectively from wastewater. However, due to the smaller size and high aggregation characteristics, it becomes difficult to separate the CNTs from the aqueous medium post-treatment. By preparing composites of other materials like metal oxide, polymers, etc. with CNTs can resolve this problem. These composites of CNT will provide a stable matrix and thereby easing the CNT separation. The creation of additional active sites is another big advantage for faster and greater dye removal capacity. In the following section, the application of some CNT nanocomposites in dye removal processes is discussed.

Chatterjee et al., 2010 studied the application of chitosan hydrogel impregnated multiwalled carbon nanotubes (MWCNTs) to remove Congo Red (CR) using a batch operation. The experimental data of the sorption of CR on prepared nanotubes fitted well to the Langmuir isotherm model with the highest adsorption uptake of 450.4 mg/g of adsorbent, which is remarkably high for such a process (Chatterjee, Lee, and Wooa 2010).

Wang et al., 2011 discussed the use of CNT-activated carbon fabric (ACF) composites in the phenol and Basic Violet 10 (BV10) dye removal from wastewater. Adsorption isotherms are characterized by Dubinin-Radushkevich and Langmuir model. The deposition of CNTs on ACF adsorbent enhanced the adsorption capacity, the adsorption energy, and the equilibrium rate constant due to the high accessibility of dye molecules to the more adsorptive sites. The maximum adsorption capacity of BV10 dye was reported to be 103.2 mg/g and 220 mg/g using Langmuir models and Dubinin-Radushkevich model respectively (Wang, Yang, and Hsieh 2011).

Bahgat et al., 2014 conducted experiments on the removal of Toluidine Blue using $NiFe_2O_4$/MWCNT composite. It was observed that the degradation rate of the dye was increased with the increase in initial dye concentration, temperature, and CNT dosage. Pseudo-second-order kinetic model and Langmuir isotherm model showed fine agreement with experimental adsorption data (Bahgat et al. 2014).

Another class of CNTs having a magnetic property is very useful in dye removal with the added advantage of easy separation from the test solution. The efficient adsorption capacity of magnetic nanoparticles is because of large specific surface area and a small resistance to diffusion (Ngomsik et al. 2005). Magnetic particles are mixed with the solution to attract the target species. Then by applying external magnetic fields, the solution and the particles were separated from each other. Various researchers reported the separation of dyes like Neutral Red, Brilliant Cresyl Blue, Methylene Blue, and Methyl Orange, using magnetic multiwall carbon nanotubes (MMWCNT) (Gong et al. 2009; Ai et al. 2011; Yu et al. 2012). MMWCNTs were prepared and utilized for removing dyes like Janus Green B, Thionine, Methylene Blue (MB), and Crystal Violet from wastewater. The effect of parameters like initial dye concentration, initial pH, contact time, and adsorbent dosage, etc. have been studied to understand the optimum adsorption and maximum separation that is possible to achieve. The maximum capacity of monolayer adsorption for Thionine, Janus Green B, Crystal Violet, and Methylene Blue dyes is reported as 36.63, 250.0, 227.7, and 48.08 mg/g respectively. Experimental data were modeled with Freundlich and Langmuir equations. Langmuir's model explains the isotherm behavior correctly. It was found that the Freundlich equation cannot predict adsorption behavior accurately. Regeneration potential is evaluated by conducting the dye desorption test where MMWCNTs was washed with acetonitrile, methanol and N, N-dimethyl formamide. The same study establishes MMWCNTs as an efficient reusable, economic sorbent where it can be reused more than 5 cycles without losing its adsorption capacity (Madrakian et al. 2011).

Various ternary nanocomposites have been developed for color removal from textile effluents. A ternary nanocomposite comprising nano-silica and xanthan gum that is grafted with hydrolyzed polyacrylamide has been synthesized to treat wastewater containing dyes such as Methyl Violet and MB. Kinetic analysis of the adsorption experiment revealed the suitability of the pseudo-second-order model for kinetic data. Langmuir equation is found to be suitable for the adsorption isotherm data. Removal efficiency and maximum amount adsorbed for MB and metyl violet dyes are reported as 99.4%; 497.5 mg/g and 99.1%; 378.8 mg/g respectively. The remarkable high adsorption capacity of the nanocomposite attributed to the H-bonding, electrostatic and dipole–dipole interactions between the molecules of the cationic dye and the adsorbent with anionic characteristics. That is why the developed nanocomposites are becoming important adsorbent for toxic dyes removal from wastewater (Ghorai et al. 2014).`

Reduced graphene-oxide/Titanium dioxide/Zinc oxide rGO/TiO_2/ZnO based ternary nanocomposite photocatalyst system was investigated to study the photocatalytic decomposition of MB dye. Experiment results show very high degradation efficiency for rGO/TiO_2/ZnO (92%) compared to that of rGO/TiO_2 (68%) and TiO_2 (47%) within 120 min of applications (Raghavan, Thangavel, and Venugopal 2015).

Zhu et al., 2010 analyzed adsorption of MO onto m-CS/γ-Fe_2O_3/MWCNTs ternary nanocomposite. By adding MWCNTs into bio-adsorbent the MO adsorption capacity can be raised by 2.2 times. Maximum sorption capacity for MO was reported as 31.52 mg/g and 9.99 mg/g with and without MWCNTs respectively. Langmuir model was the best-fitted model for isotherm data while kinetic data were well explained with the pseudo-second-order kinetic model (Zhu et al. 2010).

Long et al., 2017 developed a ternary nanocomposite based on graphene oxide/MWCNTs/Fe_3O_4, which removed heavy metal ions (Cu (II)) and low concentrated organic dye such as MB efficiently from textile wastewater. The addition of MWCNTs to the binary composite of graphene oxide and iron oxide (Fe_3O_4) enhanced the Cu (II) and MB capture from 16.67 to 20.09 mg/g and from 45.77 to 63.73 mg/g respectively. Experimental data were best fitted with the Langmuir model (Long et al. 2017).

Another ternary nanocomposite of manganese oxide/graphene oxide/polyaniline (Mn_2O_3/GO/PANI) is successfully utilized for Indigo Carmine dye removal. The adsorption capacity was found to be in the order of the combination "GO/PANI > Mn_2O_3/GO/PANI/ > PANI > PANI/MnO_2." The highest monolayer sorption capacity of 76.40 mg/g was obtained for Mn_2O_3/GO/PANI ternary composite from the Langmuir model. The process of adsorption was a spontaneous exothermic process where kinetics is described well in the pseudo-second-order model (Gemeay, Elsharkawy, and Aboelfetoh 2017).

For sonodegradation of MB, CR, MO, and RB dyes under ultrasound irradiation process, a combination of MIL-101(Cr)/RGO/$ZnFe_2O_4$ ternary magnetic nanocomposite catalyst was successfully utilized. Experiments showed the improved sonodegradation performance of nanocomposite in comparison to pure $ZnFe_2O_4$, pure MIL-101(Cr), MIL-101(Cr)/RGO. Improved degradation efficiency was achieved because of the high surface area of the nanocomposite, $ZnFe_2O_4$ magnetic characteristics along with the rapid creation and removal of charge carriers (electrons and holes) (Nirumand et al. 2018).

Polyaniline-doped with Camphor sulfonic acid i.e., PANI (CSA) -WO_3-MWCNT ternary nanocomposite has been investigated for dye removal from wastewater. The nanocomposite has been applied as a photocatalyst in MB dye degradation. The degradation rate of MB dye increased to 91.40% with the ternary nanocomposite compared to that achieved by pure PANI (CSA) (48.4%), pure WO_3 (43.45%), and PANI (CSA)-WO_3 binary nanocomposite (85.15%). Photo-catalytic and photo-electro-catalytic performance is enhanced by introducing PANI (CSA) to WO_3 because of the formation of hetero-junction (type-II) between PANI (CSA) and WO_3. On the other hand, the addition of MWCNT improves the interfacial charge separation and transferring (Hosseini et al. 2019).

Another ternary nanocomposite photocatalyst (rGO)-ZnO-TiO_2 has beens synthesized for application in Crystal Violet dye removal (Potle et al. 2020). It is observed that the photocatalytic activity of ultrasound-assisted nanocomposite is enhanced (15% higher) against the conventionally prepared nanocomposite.

7. Nano Filtration

Membrane technology in textile wastewater treatment has come into view as a feasible replacement. Nanofiltration membranes are comparatively latest development among various membrane technologies such as reverse osmosis, microfiltration, and ultrafiltration membranes. Here the process of separation is characterized by an organic and composite thin-film membrane with pore sizes in 0.1-10 nm range. Nanofiltration membranes have the ability to function at low pressure and to reject solute selectively based on the size and charge which is not exhibited by the reverse osmosis (RO) membranes. This is not possible in reverse osmosis membranes that discard all solutes. Generally, the setup cost of most of the filtration techniques is quite large but it gets compensate by reusing salts and permeates. By regularly washing membrane to avoid fouling issues, using pre-filters, and by selecting the most suitable membrane system, the process can also be made economic and profitable.

A positively charged Nanofiltration membrane was fabricated to treat textile dye effluent and optimized the process w.r.t. various parameters as operating pressure and feed flow. Studies showed that an applied pressure of 1.0 MPa and a feed flow of 40 liters/hour (l/h) helped the membrane in achieving 93.3% COD removal and reducing around 51.0% in TDS, salinity, and conductivity along with achieving 100% Chroma removal. Also, the membrane could be recharged by cleaning with a 1% NaOH solution (Huang et al. 2012).

Submerged nanofiltration (SNF) was found to possess high water cleaning efficiency and the performance was studied by utilizing a hollow fiber nanofiltration membrane fabricated in the laboratory. Both color and COD from already biologically treated effluent can be effectively eliminated using a nanofiltration membrane whose molecular weight cut off is around 620 Dalton (Da). A decline in water permeability, COD level, and a rise in color degradation rate was observed with an increase in trans-membrane pressures (TMP) and/or volume concentrating factors (VCF). The submerged nanofiltration system attained a steady flux of 5.15 l/m^2h, 99.3% color degradation rate, and 91.5% COD reduction rate under the optimum conditions of TMP (0.8 bar) and VCF (4.0) (Zheng, Yu, Shuai, Zhou, Cheng, et al. 2013).

A nanofiltration membrane (average molecular weight 22,400 Da) was synthesized by phase inversion technique with Polysulfone using as the base polymer and N, N-dimethyl formamide (DMF) using as a solvent. Textile effluents containing three anionic reactive dyes such as Basic Blue, Cibacron Black, Cibacron Red, and Acidic Yellow were passed through the membrane in two stages at varying TMP and cross-flow rate (CFR). The results were studied on the basis of salt recovery, COD, and dye rejection. Separation of dye was largely due to size exclusion facilitated by charge-charge interaction. Around 99% of the dye is rejected, 90% COD level is reduced, and up to 70% of salt is recovered after the second stage filtration. Almost complete flux recovery in the second stage confirms long membrane life. To restore the membrane permeability, it was washed with tap water and then distilled water at 966 kPa trans-membrane pressure and 120 l/h crossflow rate (Panda and De 2015).

Polyamide-imide hollow fiber Nanofiltration (NF) membrane was used at lab scale as well as pilot-scale and the performance was evaluated in treating textile wastewater under various operating conditions (feed temperature, solute concentration, and pH). At nearly all testing conditions, the NF membrane showed an average of more than 90% rejection against several dyes. Around 80% of NaCl and 90% of Na_2SO_4 can diffuse through the membrane, thereby, the membrane provided the high potential to recover and reuse these salts in the subsequent dyeing process (Ong et al. 2014).

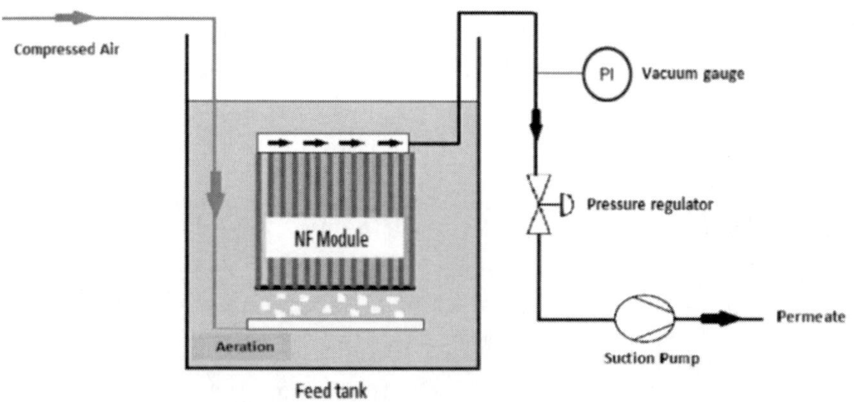

Figure 1. Submerged Nanofiltration experimental set up. Modified after (Zheng, Yu, Shuai, Zhou, and Cheng 2013).

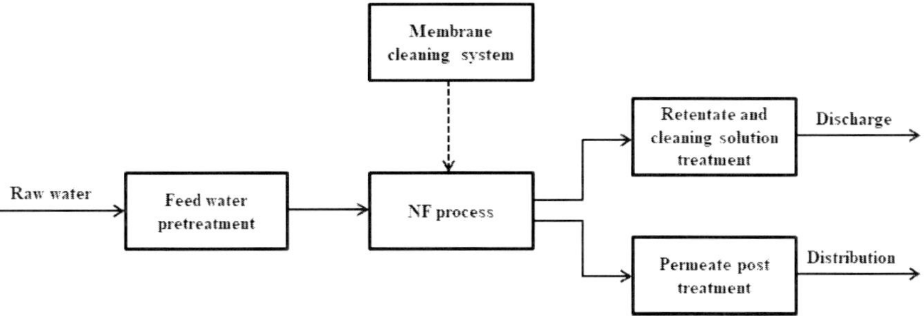

Figure 2. Water Nanofiltration process. Modified after (Khosravi et al. 2020).

Polysulfone based hollow fiber nanofiltration membrane was prepared by interfacial polymerization of meta-phenylenediamine and trimethyl chloride (TMC). At TMC concentration 1 wt%, the membrane had the molecular weight cut off of 360 Da with a permeability of $1.14x10^{11}$ m/Pa-s and an average pore radius of 5 Å. This membrane showed 99.9% rejection of Reactive black, red and yellow dyes, 98.8% rejection of blue dye, present in textile waste effluent at 104 kPa TMP, and 30 l/h CFR along with 42% salt recovery (Mondal and De 2016).

Studies showed that nanofiltration using a spiral-wound module is also an effective way to remove dyes such as methylene blue, reactive blue 4, and acid orange 10 from the wastewater. The module consists of a centrally pierced tube around which the membrane is wounded. Among all the dyes studied, the membrane rejected the Reactive Blue 4 dye in maximum quantity possibly due to its highest molecular weight and less solubility compared to other dyes. In the spiral wound module, the feed was injected in the axial direction of the module, and permeate were collected in the central tube after traversing the membrane leaves. A Spiegler-Kedem based model predicted the characteristics of the permeate without much deviation (Dash and Kumar 2017).

Efforts were made to convert commercial Polyethersulfone ultrafiltration membrane to Nanofiltration membranes by the coating of poly (vinyl alcohol) (PVA) using the dip-coating method. To enhance the cross-linking PVA on the surface, malic acid was used. The evaluation of membrane performances was done using cross-flow filtration of dispersing, reactive and acid dyes. It was found that membrane with 1 wt. % PVA obtained the flux of 3.06 l/m^2 h atm. An increase in pressure lead to more rejection of dyes initially but after reaching the maxima, the rejection of dyes decreases. The highest rejection was found to be 97.8% at 15 atm for dispersed dyes, whereas the maximum rejection value was found to be 95.7% and 93.5% for the remaining dyes. To study mass transfer coefficients and solute transport parameters combined film-theory solution-diffusion model was used (Babu and Murthy 2017).

From the co-deposition of ε-Poly-L-lysine and pyrogallic on the hydrolyzed polyacetonitile substrate, another versatile loose composite NF membrane was developed and then for the ε-PL/PG mediated surface, PEGylation was done. The prepared membrane possessed eminent bactericidal property against E. coli and S. aureus. On the treatment of textile wastewater, the membrane showed a high permeability (76.3 l/m^2 h MPa), a high removal rate of COD (89.1%), and a high rejection for dye Methyl Blue (98.9%) with a low rejection of NaCl (17%). The flux recovery ratio was 85%, which indicated the satisfactory antifouling performance of the membrane in the long-term treatment for textile wastewater (S. Zhao et al. 2018).

Instead of using individual nanofiltration, if the process was used in combination with other processes, the overall results were improved. Research showed that if the textile wastewater was treated with microfiltration–membrane bioreactor–nanofiltration hybrid process, the results were even better. The optimal TMP value was determined to be 12 bars, at which good permeate characteristics were obtained with reasonable energy consumption (Couto et al. 2017).

In another study, Tavangar et al. utilized electrocoagulation (EC) and nanofiltration (NF) individually as well as their combination (EC-NF) to treat wastewater and to compare the process. In EC pretreatment, various electrodes such as Ti, Fe, and Al were utilized. Results showed that among all electrodes around 93% color removal, 64% reduction in COD level, and 99% reduction in turbidity was achieved using the Al electrode which proved to be the optimal EC electrode. Better results were obtained using loose NF membrane (NP010, Microdyn Nadir) with the removal of 87% color, 74% COD, 99% turbidity. The amalgamation of EC-NF displayed extraordinary performance and removed their individual faults. From the EC-treated residual, the NF membrane is capable of eliminating the highly intense color to enhance the overall performance. Also, membrane fouling was reduced, and the permeate flux of the NF membrane increased from EC pretreatment (Tavangar et al. 2019).

If biological pretreatment is coupled with an NF membrane, the process becomes much more effective for water reclamation. Anaerobic phase and aerobic phase (sequencing batch reactor) are the components of biological pretreatment, which helps in structural degradation of dye molecules and also reduces COD concentration. The pretreatment of the membranes is done to make them more efficient and to achieve high-quality effluents.

For removing Reactive Blue 21 (RB21), a highly soluble phthalocyanine dye, and sodium dodecyl sulfate (SDS), a commercial anionic surfactant from the industry effluent, a sequencing batch reactor/NF hybrid method was studied and concluded to be an effective method. However, no individual treatment method can individually retrieve water from textile wastewater. In the sequencing batch reactor section, the optimum reaction time was reported to be in a ratio of 8 h anaerobic/3 h aerobic. The outflow of sequencing batch reactor was used as an NF membrane feed in NF treatment. The above method eliminated dye, COD, and SDS with an efficiency of 98%, 98.5%, and 99%, respectively (Khosravi et al. 2020).

8. Nano Catalyst

In a catalysis process, the rate of a reaction is enhanced with the adding of a foreign substance known as a catalyst. It is well known to be categorized as homogenous and heterogeneous catalysis. In a way, both of these catalysis has its advantages and disadvantages. Homogenous catalysis is active whereas, in heterogeneous catalysis, a catalyst is easily recoverable.

Nano catalytic system has high selectivity and it can also be easily recovered. A large surface area is available for the reactants to come into the catalyst contact. It is also insoluble in the reaction solvent similar to heterogeneous catalysis (Tandon 2014).

Various catalysts are used in textile industries in order to accelerate chemical reactions and to operate in milder conditions. It helps in replacing harmful chemicals, it is easy to control and it is also recyclable. In the textile industry, catalysts are used in dyeing, finishing, and effluent treatments.

Generally, metal and metal oxides are used as heterogeneous catalysts. But it is hardly used in dyeing because they increase the hardness of the water which in turn leads to precipitation of dyes. Metal and their oxides can be used as a catalyst in wastewater treatment because of the insoluble nature of vat and sulfur dyes. Metal oxide nanoparticles are also used in textile finishing enhancing performance and adding special functional properties to the fabrics (Lam, Kan, and Yuen 2012).

8.1. Nano Catalysts in Photocatalytic Degradation of Dyes

Over the past few years, photocatalysis has emerged as a promising and efficient advance oxidation process (AOP) for removing dyes contaminants from textile industries. Photocatalytic degradation involves heterogeneous catalysis. It involves the use of a semiconducting material that can be excited by the absorption of light with the generation of hydroxyl radicals from the metal oxide nanoparticles (NPs). These radicals act as a strong oxidant and break dye molecules into simpler molecules and finally convert into carbon dioxide and water (Reza, Kurny, and Gulshan 2017).

Photocatalytic degradation of dyes by metal oxide nanoparticles (NPs) shows many advantages over conventional treatment methods. It has inherent destructive nature, it can be carried out at low temperature and pressure, highly efficient, and the complete degradation of dyes can occur within a few hours under ambient conditions. The ultimate products of the degradation of the organic pollutants are CO_2 and water without the formation of secondary harmful material. In recent years, various metal oxides nanoparticles like TiO_2, ZnO, and other oxides have gained attention for photocatalytic degradation of dyes. Among all the metal oxide nanoparticles (NPs), TiO_2 has proved itself and gained attention to be the best material due to its low cost, non-toxicity, high physical and chemical stability, easy availability, high photocatalytic activity, high water solubility and high reactivity (Mortazavian, Saber, and James 2019).

The complete degradation process of dyes using photocatalysts is shown in Figure 3. On illuminating photocatalyst with photons of energy that is equivalent or larger than its bandgap width, excitation of electrons to the conduction band (CB) of the photocatalyst takes place creating holes in the valence band (VB). The generated holes and the electrons then diffuse to the photocatalyst surface and react with the adsorbed dye molecule. These electrons and holes then produce radical species, such as hydroxyl radicals ($^{\cdot}OH$) and superoxide radicals ($^{\cdot}O_2^-$) respectively. These reactive species rapidly and non-selectively degrade dye molecules. The complete photocatalytic degradation process depends upon some basic parameters such as the catalyst dosage, pH, concentration of dye, and light intensity (Chiu et al. 2019). The photocatalytic degradation mechanism described above can be summarized as by taking an example of TiO_2 metal oxide (Min et al. 2015):

$$TiO_2 + h\nu\ (UV) \rightarrow TiO_2\ [e^- + h^+] \tag{1}$$

$$H_2O\ (ads) + h^+ \rightarrow OH^{\cdot}(ads) + H^+\ (ads) \tag{2}$$

$$O_2 + e^- \rightarrow O_2^-\ (ads) \tag{3}$$

$$O_2^-\ (ads) + H^+ \leftrightarrow HOO^{\cdot}(ads) \tag{4}$$

$$2HOO^{\cdot}(ads) \rightarrow H_2O_2\ (ads) + O_2 \tag{5}$$

$$H_2O_2\ (ads) \rightarrow 2OH^{\cdot}(ads) \tag{6}$$

$$Dye + OH^{\cdot} \rightarrow CO_2 + H_2O \tag{7}$$

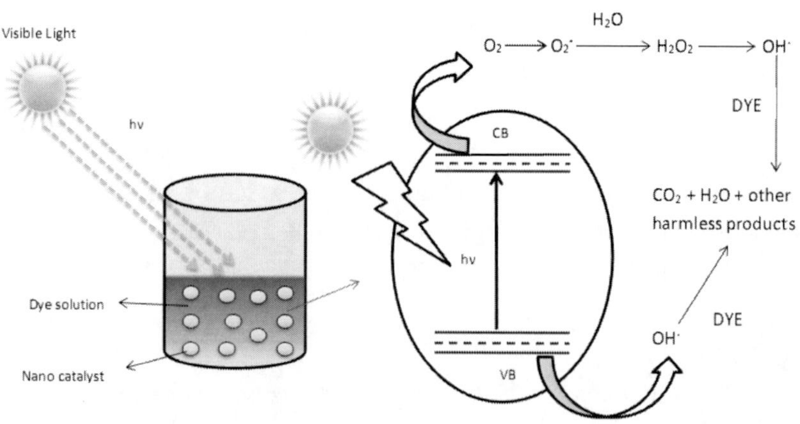

Figure 3. Photocatalytic process of dye removal. Modified after (Min et al. 2015).

Xiaoqing Chen et al., 2017 prepared ZnO photocatalyst using zinc acetate as a precursor by sol-gel method and was employed in photodegradation of azo dyes such as Direct Red 38 (DR38), MO, and CR under UV irradiation. Initial concentrations of MO, CR, and DR38 were varied from 10 to 50 mg/l to study the effect of concentration of dyes on photocatalytic degradation performance keeping the constant value of irradiation time 10 minutes, catalyst amount 0.2 g/l and pH 6.8. On increasing the initial azo dye concentration, it was noticed that the degradation efficiency notably decreased from 99.53% to 56.61%, 99.14% to 48.03%, and 99.65% to 40.64% for MO, CR, and DB38 respectively. Also, the effect of catalyst dosage ranging from 0.1 to 0.8 g/l on the degradation of azo dyes was also examined keeping initial dye concentration 30 mg/l, irradiation time 10 min, and pH 6.8 constant. The rate of degradation raised from 68.0 to 99.7%, 56.49 to 99.21%, and 49.25 to 99.45% for MO, CR, and DB38 respectively. The first-order kinetic model best fitted the experimental data (X. Chen et al. 2017).

Stoyanova et al., 2018 synthesized lanthanum (La) modified TiO_2 photocatalyst by nonhydrolytic sol-gel method. The effect of pH on degradation efficiency of dyes under UV light was studied by varying pH values from 3.9 to 9.8 by adding 1 mol solution of HCl or NaOH. The initial dye concentration (20 mg/l RB5 and 14 mg/l CR) and La-TiO_2 catalyst amount (0.67 mg/ml) were kept constant. It was observed that for both the dyes, maximum degradation efficiency was obtained at their natural pH values, pH 6.9 for RB5, and pH 6.1 for CR. The effect of catalyst amount on decoloration rate of RB5 and CR was studied keeping initial dye concentration (14 mg/l CR and 20 mg/l RB5) and pH (6.1 CR and 6.9 RB5) constant. It was noticed that on increasing the amount of photocatalyst, decoloration of both the dyes increases reaching the peak decoloration for a catalyst amount of 1 mg/ml. After that, it starts to decrease with an increase in the photocatalyst amount (Stoyanova, Bachvarova-Nedelcheva, and Iordanova 2018).

Saeed et al., 2019 prepared multiwalled nanotubes/copper and titanium oxides (MWNTs/Cu-Ti) composite by wet chemical precipitation technique. Various experiments were performed under UV-light irradiation to study the photodegradation activity of alizarin red (AR) dye. Experimental results showed that in 0.5 h about 22.75% of the dye was degraded, which rose to 92.51%, on increasing irradiation time to 3 hours. Different amounts of catalyst (0.015, 0.020, 0.025, 0.030, 0.035, 0.040, and 0.045) were charged for studying the effect of catalyst dosage on the photodegradation rate by keeping dye concentration 200 ppm and irradiation time 1 hour constant. On increasing the dosage of the catalyst, it was observed that the photodegradation rate of dye also increased up to an optimum level. Dye concentration was changed from 100 to 300 ppm to study the photodegradation rate of AR with a constant irradiation time of 1 hour and a catalyst amount of 0.02 g. It was noted that on increasing initial dye concentration, the degradation rate decreased. Also, pH values were varied from 3 to 11 with an irradiation time of 1 hour, dye concentration of 200 ppm, and a catalyst amount of 0.02 g. On increasing pH values, it was noticed that the degradation rate also increased. The results showed that at pH 3 about 85.82% of dyes degraded, at pH 7 about 92.51% of dye degraded, and 96.61% of dye degraded at pH 11 (Saeed, Zada, and Khan 2019).

Kiwaan et al. in 2020 synthesized TiO_2 photocatalyst using $TiCl_4$ as a precursor by low-temperature co-precipitation method. To examine the photocatalytic activity of the prepared catalyst, different calcination temperature was set. It was noticed that on increasing the calcination temperature, the degradation rate of Acid Red 57 (AR57) and Rhodamine B (RB) decreased. About 93.8% RB dye and 90.7% AR57 dye removed at a calcination temperature of 673K. Also, keeping constant catalyst concentration of 2.5 g/l and pH 7, dye concentration was varied from 4-9 mg/l for RB and 10-50 mg/l for AR57. On increasing the dye concentration, it was found that the degradation rate increased at first up to 6 mg/l for Rhodamine B and 30 mg/l for Acid red 57 and then decreased. The catalyst amount was also differed from 0.1 to 0.5 g/l to study the effect of catalyst dosage on the degradation rate with a constant initial dye concentration of 6 mg/l for RB and 30 mg/l for AR57, the irradiation time of 20 min and pH 7. With a catalyst amount of 2.5 g/l, the highest photocatalytic degradation efficiency was achieved for both the dyes, and then a sudden decrease was observed with a further increment of catalyst amount. Adsorption data were fitted well with the Langmuir isotherm and first-order kinetics (Kiwaan et al. 2020).

9. Electrospun Nanofibrous Membranes

Textile industries eject large quantities of wastewater into freshwater bodies containing various polluting dyes. Membrane technologies have been playing an important role in wastewater treatment. Over the past few years, nanofibers are gaining great attention in their application to remove dye contaminants because of the extraordinary properties of these materials. The special advantages of membrane processes as compared to other treatment methods are: (1) less power consumption (2) lenient operating conditions (3) small footprint. The various techniques for polymer membranes fabrication are electrospinning, phase separation, centrifugal spinning, drawing, self-assembly, template synthesis, melt blowing, and stretching, track-etching, sintering, solvent evaporation (Asad, Sameoto, and Sadrzadeh 2020). Among all the above methods, over the past decade, electrospinning has become an eminent method to fabricate polymeric nanofibrous membranes because of its direct and resilient methodology to fabricate nanofibers economically from lab scale to pilot and industrial scales in different sizes and shapes. The *"Electrospun Nanofibrous Membranes"* have unique and interesting features like high adsorption capacity, low secondary pollution, high porosity, high surface area to volume ratio, and good mechanical properties (Nasreen et al. 2013).

The electrospinning process contains four major parts: a syringe needle, a syringe pump, a collector, and a high-voltage source or supplier. First, a polymer solution should be prepared by dissolving a polymer into an appropriate solvent. Once the polymer solution dissolves completely, it is then introduced into the syringe pump for fabricating the fibers. To trigger the jet towards the collector, a high voltage source is used. The electrical conductivity of the solution kept high to increase the electrospinnability. In this process, between the syringe containing the solution and a collector, high voltage is applied. A Taylor cone appears and it forms a liquid jet and travels towards a collector plate when the voltage is increased gradually and when the applied voltage exceeds the surface, the solvent evaporates and fibers deposits on the collector surface. The complete process diagram is shown in figure 4.

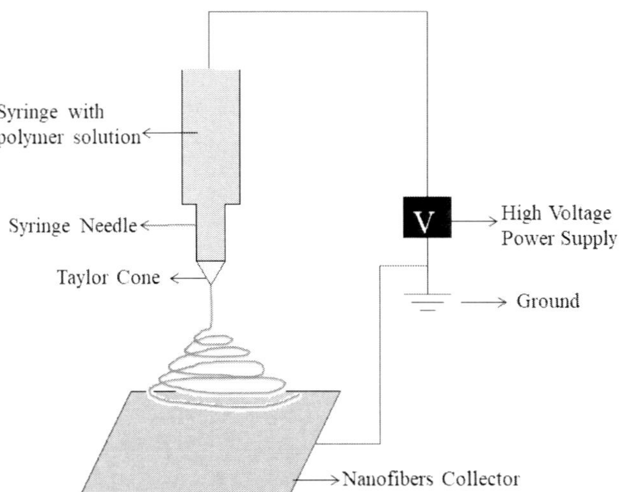

Figure 4. Schematic diagram of the electrospinning process. Modified after (Nasreen et al. 2013).

The structure and morphology of the electrospun nanofibrous membranes depend upon several factors divided into three main categories: solution parameters (concentration, viscosity, surface tension, conductivity, and polymer molecular weight), processing parameters (tip to collector distance, applied voltage, feed rate, type of collector) and ambient parameters (humidity, temperature) (Suja et al. 2017). Table 4 shows a brief description of the parameters and their effects on fiber morphology and structure.

Table 4. Various parameters affecting electro spun Nanofiber morphology and structure. Modified after (Hasanzadeh and Moghadam 2013)

Parameters	Description	Effect on nanofibers morphology
	Solution parameters	
Concentration	Increment in concentration	Increment in fiber diameter
Polymer molecular weight	Increment in molecular weight	Nanofibers with fewer beads
Viscosity	Increment in viscosity	Thicker and beadles nanofibers
	Very high viscosity	Beads generation
Surface tension	Increment in surface tension	No effect on nanofibers morphology
Conductivity	Increment in conductivity	The decrement in diameter of the fiber
	Processing Parameters	
Applied voltage	Increment in voltage	The decrement in the fiber diameter
Tip to collector distance	Very low distance	Formation of bead
	Too large distance	
Feed rate	Very low flow rate	The decrement in fiber diameter
	Too high flow rate	Formation of beads
Type of collector	Metal collectors	Preferred
	Conductive collectors	Aligned nanofibers are obtained
	Ambient parameters	
Temperature	Increment in temperature	The decrement in fiber diameter
Humidity	High humidity	Circular pore formation on the fiber
	Low humidity	Less bead formation

Ghani et al., 2016 fabricated Alg/PEO (Alginate/Polyethylene Oxide) nanofibrous membranes by electrospinning method. In the electrospinning set up, 12 KV voltage source was used, rolled aluminum foil was used as a collector which is maintained at 20 rpm, distance from the tip of the needle to the collector was 10 cm and feed rate was maintained at 8 ml/h. pH values were varied to examine the dependence of pH on dye adsorption with a constant dye amount of 50 mg/l and the membrane dosage of 0.5 g. It was observed that dye degradation efficiency increased on decreasing the pH values for anionic Acid Red 14 (AR14) and vice versa for cationic Basic Blue 41 (BB41) dye. The results showed that at pH of 1 and 9, the maximum adsorption uptake for AR14 and BB41 was 93% and 71% respectively. Also, different amounts of the membrane were added to dye solution to understand the effect of membrane dosage on the adsorption of two dyes keeping the concentration constant at 50 mg/l and constant pH at 1 and 9 for AR41 and BB41 respectively. On increasing the membrane dosage, dye removal efficiency for both the dyes increased. The optimum value of membrane dosage was observed as 0.2 g. On further increasing the dosage of the membrane, the adsorption capacity decreased because of the blocking of some adsorption sites on the membrane (Ghani et al. 2016).

Aziz et al. in 2017 prepared silk fibroin (SF)/polyacrylonitrile (PAN) double layer nanofibrous membranes containing different compositions (0, 2.5, 5, 7.5 wt.%) of polyaniline/TiO_2 nanoparticles using electrospinning method. In the electrospinning process, 16KV and 22KV high voltage sources were applied, the feed rate was maintained at 1 ml/h, aluminum foil was utilized as a collector and distance from collector to the tip of the needle was maintained as 14 cm. The effect of the addition of polyaniline (PANI)/TiO_2 nanoparticles on adsorption of anionic reactive black HFGR dye on nanofibrous membranes was studied. It was observed that on increasing the amount of nanoparticles in the nanofilters, the adsorption capacity of nanofibrous membranes has increased. The dye degradation rate increased from 20.2% to 61.6% at a dye concentration of 100 ppm for the addition of 7.5 wt.% nanoparticles within the prepared nanofibrous membranes. Also, on increasing the dye concentration in the solution, a small decrease in dye removal efficiency was observed. Also, for pH values of 3, 7.8 and 10, and dye concentration of 50 and 75 ppm at room temperature, the effect of pH on adsorption of HFGR dye on SF-PAN nanofilters with 7.5 wt.% PANI/TiO2 nanoparticles was examined. It was noted that around 92% degradation rate was obtained at a dye concentration of 50 ppm and pH value 3 (Aziz et al. 2017).

J. Fendi and A. Naser et al., in 2018 prepared two electrospun nanofibrous membranes. The first membrane (M1) was prepared form polymer solution of p-cresol formaldehyde and polystyrene dissolved in dimethyl formaldehyde as a solvent. The second membrane (M2) was prepared by doping the first solution with zinc oxide nanoparticles in a ratio of 3% w/v. The electrospinning device was set at 30 rpm for a cylindrical collector with aluminum foil wrapping, 40KV voltage. With a dye concentration of 18 mg/l at 298K, the effect of membrane dosage and contact time on adsorption was studied for both the membranes. As the contact time increased, it was observed that adsorption capacity also increased up to a certain value after 90 min. Also, on increasing the dosage of the adsorbent membrane, a slight increase in the adsorption was noticed. The highest value observed was 0.015 g, after that no change was noticed. On changing initial concentrations of MB dye (3, 6, 9, 12, 15, and 18 mg/l); it was observed that adsorption uptake decreased with increasing dye concentration for both membranes. Again, by changing the temperature (298, 303, 308, 313, and 318K) keeping initial concentration (18 mg/l and 0.015 g of both membranes) constant, it was founded that the adsorption uptake increased on increasing the temperature. It was also observed that adsorption follows the Langmuir isotherm model (J. Fendi and A. Naser 2018).

Xu et al., 2019 fabricated electrospun polyethersulfone nanofibrous membranes by a one-step electrospinning method. In the electrospinning set up, 14KV voltage was applied, the feed rate was kept at 0.5 ml/min and aluminum foil was used as a collector. To study the effect of solution pH on adsorption, pH values were varied from 2 to 9. On raising pH values, the MB dye degradation rate increased up to a certain pH and then the degradation rate decreased between pH 9 and 11. At pH 9, the highest degradation rate was attained. PSO kinetic model was found to be the best fit for experimental adsorption kinetics data and the highest adsorption uptake of 2257.88 mg/l was calculated from the Langmuir isotherm model (Xu et al. 2019).

Cheng et al., 2020 fabricated deacetylated cellulose acetate (DA) @ polydopamine (PDA) composite nanofibrous membranes using the electrospinning method. It was found that the rate of adsorption increased remarkably and then became stable after some time. At the starting of adsorption, the possibility of contact between the MB dye molecules and the DA@PDA nanofibers membrane was high since dye concentration was high. After 24h, the adsorption attained a steady-state, and the adsorption capacities were found to be 4.9, 10.2, and 88.2 mg/g for cellulose acetate, DA, and DA@PDA respectively. Also, on increasing pH from 2 to 10, it was noticed that the adsorption uptake of the DA@PDA membrane significantly increased from 1.43 to 92.64 mg/g. At pH 10, the dye removal efficiency was noticed as 93.21%. The experimental adsorption isotherm data was found to be best fitted with the Langmuir isotherm model whereas the kinetic data were modeled with the pseudo-second-order kinetic equation (Cheng et al. 2020).

CONCLUSION

Dye removal is essential for all the textile industries to maintain a healthy and sustainable ecosystem. Nanotechnology-based materials present potential processes for efficient dye removal from textile effluent. A variety of materials including metal oxides, alloys, composites, CNTs, both obtained from chemical as well as biological resources at the nanoscale have been successfully used for this purpose, as is clearly depicted by the removal efficiency and kinetic studies. However, few challenges have always been experienced during the process, such as the successful recovery of the material after dye removal, the cost-effectiveness of the process, scaling up the processes, etc. Although various researchers have made use of nanotechnology in different forms for dye removal in textile industries, the studies have been found applicable only at the lab scale. The need and scope are there to scale up the process and commercialize it, and would even be more appealing, if the material developed and optimized for this purpose finds biological origin, which would help in reducing pressure on petrochemical industries, along with value addition to the waste generated in nature.

NOMENCLATURE

Parameter	Description
mg	Milligram
g	Gram
l	Liter
h	Hour
m^2	Square meter
Nm	Nanometer
bar	Pressure in bar unit
kPa	Pressure in kilo Pascal unit
Å	Angstrom
M	Meter
s	Second
atm	Pressure in atmospheric (atm.) unit
ml	Milliliter
ppm	Parts per million
K	Temperature in kelvin scale
KV	Kilo Volt
rpm	Revolution per minute

REFERENCES

Agboola, Oluranti, Patricia Popoola, Rotimi Sadiku, Samuel Eshorame Sanni, Sunday Ojo Fayomi, and Olawale Samuel Fatoba. 2018. "Nanotechnology in Wastewater and the Capacity of Nanotechnology for Sustainability." In *Environmental Nanotechnology Volume 3*, 1–45.

Ai, Lunhong, Chunying Zhang, Fang Liao, Yao Wang, Ming Li, Lanying Meng, and Jing Jiang. 2011. "Removal of Methylene Blue from Aqueous Solution with Magnetite Loaded Multi-Wall Carbon Nanotube: Kinetic, Isotherm and Mechanism Analysis." *Journal of Hazardous Materials* 198 (December): 282–90. https://doi.org/10.1016/j.jhazmat.2011.10.041.

Allen, N. S. 1971. *Colour Chemistry*. https://doi.org/10.1016/1010-6030(87)80016-3.

Amin, M. T., A. A. Alazba, and U. Manzoor. 2014. “A Review of Removal of Pollutants from Water/Wastewater Using Different Types of Nanomaterials.” *Advances in Materials Science and Engineering* 2014. https://doi.org/10.1155/2014/825910.

Asad, Asad, Dan Sameoto, and Mohtada Sadrzadeh. 2020. “Overview of Membrane Technology.” *Nanocomposite Membranes for Water and Gas Separation*, 1–28. https://doi.org/10.1016/b978-0-12-816710-6.00001-8.

Aziz, Soroush, Mohammad Sabzi, Ali Fattahi, and Elham Arkan. 2017. “Electrospun Silk Fibroin/PAN Double-Layer Nanofibrous Membranes Containing Polyaniline/TiO2 Nanoparticles for Anionic Dye Removal.” *Journal of Polymer Research* 24 (9): 140. https://doi.org/10.1007/s10965-017-1298-0.

Babu, Jibin, and Z. V. P. Murthy. 2017. “Treatment of Textile Dyes Containing Wastewaters with PES/PVA Thin Film Composite Nanofiltration Membranes.” *Separation and Purification Technology* 183: 66–72. https://doi.org/10.1016/j.seppur.2017.04.002.

Bahgat, M., A. A. Farghali, W. M.A. El Rouby, and M. H. Khedr. 2014. “Efficiency, Kinetics and Thermodynamics of Toluidine Blue Dye Removal from Aqueous Solution Using MWCNTs Decorated with NiFe2O4.” *Fullerenes Nanotubes and Carbon Nanostructures* 22 (5): 454–70. https://doi.org/10.1080/1536383X.2012.684188.

Cao, Xue Li, Ya Nan Yan, Fu Yi Zhou, and Shi Peng Sun. 2020. “Tailoring Nanofiltration Membranes for Effective Removing Dye Intermediates in Complex Dye-Wastewater.” *Journal of Membrane Science* 595 (September 2019). https://doi.org/10.1016/j.memsci. 2019.117476.

Chatterjee, Sudipta, Min W. Lee, and Seung H. Wooa. 2010. “Adsorption of Congo Red by Chitosan Hydrogel Beads Impregnated with Carbon Nanotubes.” *Bioresource Technology* 101 (6): 1800–1806. https://doi.org/10.1016/j.biortech.2009.10.051.

Chen, Xiaoqing, Zhansheng Wu, Dandan Liu, and Zhenzhen Gao. 2017. “Preparation of ZnO Photocatalyst for the Efficient and Rapid

Photocatalytic Degradation of Azo Dyes." *Nanoscale Research Letters* 12 (1): 143. https://doi.org/10.1186/s11671-017-1904-4.

Chen, Y. H. 2011. "Synthesis, Characterization and Dye Adsorption of Ilmenite Nanoparticles." *Journal of Non-Crystalline Solids* 357 (1): 136–39. https://doi.org/10.1016/j.jnoncrysol.2010.09.070.

Cheng, Jiaqi, Conghua Zhan, Jiahui Wu, Zhixiang Cui, Junhui Si, Qianting Wang, Xiangfang Peng, and Lih Sheng Turng. 2020. "Highly Efficient Removal of Methylene Blue Dye from an Aqueous Solution Using Cellulose Acetate Nanofibrous Membranes Modified by Polydopamine." *ACS Omega* 5 (10): 5389–5400. https://doi.org/10.1021/acsomega.9b04425.

Chiu, Yi-Hsuan, Tso-Fu Mark Chang, Chun-Yi Chen, Masato Sone, and Yung-Jung Hsu. 2019. "Mechanistic Insights into Photodegradation of Organic Dyes Using Heterostructure Photocatalysts." *Catalysts* 9 (5): 430. https://doi.org/10.3390/catal9050430.

Couto, Carolina Fonseca, Larissa Silva Marques, Míriam Cristina Amaral Santos, and Wagner Guadagnin Moravia. 2017. "Coupling of Nanofiltration with Microfiltration and Membrane Bioreactor for Textile Effluent Reclamation." *Separation Science and Technology* 6395 (April). https://doi.org/10.1080/01496395.2017.1321670.

Crini, Grégorio, Eric Lichtfouse, Lee D. Wilson, and Nadia Morin-Crini. 2019. "Conventional and Non-Conventional Adsorbents for Wastewater Treatment." *Environmental Chemistry Letters* 17 (1): 195–213. https://doi.org/10.1007/s10311-018-0786-8.

Dalvand, Arash, Ramin Nabizadeh, Mohammad Reza Ganjali, Mehdi Khoobi, Shahrokh Nazmara, and Amir Hossein Mahvi. 2016. "Modeling of Reactive Blue 19 Azo Dye Removal from Colored Textile Wastewater Using L-Arginine-Functionalized Fe3O4 Nanoparticles: Optimization, Reusability, Kinetic and Equilibrium Studies." *Journal of Magnetism and Magnetic Materials* 404: 179–89. https://doi.org/10.1016/j.jmmm.2015.12.040.

Dash, Bibek, and Abhishek Kumar. 2017. "Nanofiltration for Textile Dye–Water Treatment: Experimental and Parameter Estimation Studies Using a Spiral Wound Module and Validation of the Spiegler–Kedem-

Based Model." *Separation Science and Technology (Philadelphia)* 52 (7): 1216–24. https://doi.org/10.1080/01496395.2017.1282965.

Drumond Chequer, Farah Maria, Gisele Augusto Rodrigues de Oliveira, Elisa Raquel Anastacio Ferraz, Juliano Carvalho, Maria Valnice Boldrin Zanoni, and Danielle Palma de Oliveir. 2013. "Textile Dyes: Dyeing Process and Environmental Impact." *Eco-Friendly Textile Dyeing and Finishing*. https://doi.org/10.5772/53659.

Ejder-Korucu, Mehtap, Ahmet Gürses, Çetin Dogar, Sanjay K. Sharma, and Metin Açikyildiz. 2015. "Removal of Organic Dyes from Industrial Effluents: An Overview of Physical and Biotechnological Applications." In *Green Chemistry for Dyes Removal from Waste Water: Research Trends and Applications*, 1–34. https://doi.org/10.1002/9781118721001.ch1.

El-Aassar, M. R., M. F. El-Kady, H. Shokry Hassan, and Salem S. Al-Deyab. 2016. "Synthesis and Characterization of Surface Modified Electrospun Poly (Acrylonitrile-Co-Styrene) Nanofibers for Dye Decolorization." *Journal of the Taiwan Institute of Chemical Engineers* 58: 274–82. https://doi.org/10.1016/j.jtice.2015.05.042.

Gemeay, Ali H, Rehab G Elsharkawy, and Eman F Aboelfetoh. 2017. "Graphene Oxide/Polyaniline/Manganese Oxide Ternary Nanocomposites, Facile Synthesis, Characterization, and Application for Indigo Carmine Removal." *Journal of Polymers and the Environment* 0. https://doi.org/10.1007/s10924-017-0947-z.

Ghani, Mozhdeh, Babak Rezaei, Aliakbar Ghare Aghaji, and Mokhtar Arami. 2016. "Novel Cross-Linked Superfine Alginate-Based Nanofibers: Fabrication, Characterization, and Their Use in the Adsorption of Cationic and Anionic Dyes." *Advances in Polymer Technology* 35 (4): 428–38. https://doi.org/10.1002/adv.21569.

Ghorai, Soumitra, Asish Sarkar, Mohammad Raou, Asit Baran Panda, Holger Scho, and Sagar Pal. 2014. "Enhanced Removal of Methylene Blue and Methyl Violet Dyes from Aqueous Solution Using a Nanocomposite of Hydrolyzed Polyacrylamide Grafted Xanthan Gum and Incorporated Nanosilica." *American Chemical Society* 7: 4766–77.

Girgis, Emad Azmy, and Christen Tharwat Aziz. 2013. The use of Nano alloys in wastewater Treatment, issued 2013.

Gong, Ji Lai, Bin Wang, Guang Ming Zeng, Chun Ping Yang, Cheng Gang Niu, Qiu Ya Niu, Wen Jin Zhou, and Yi Liang. 2009. "Removal of Cationic Dyes from Aqueous Solution Using Magnetic Multi-Wall Carbon Nanotube Nanocomposite as Adsorbent." *Journal of Hazardous Materials* 164 (2–3): 1517–22. https://doi.org/10.1016/j. jhazmat. 2008.09.072.

Hasanzadeh, M, and B Hadavi Moghadam. 2013. "Electrospun Nanofibrous Membranes as Potential Adsorbents for Textile Dye Removal-A Review." *Journal of Chemical Health Risks* 3 (2): 15–26.

Hosseini, Mir Ghasem, Pariya Yardani Sefidi, Ahmet Musap Mert, and Solen Kinayyigit. 2019. "Investigation of Solar-Induced Photoelectrochemical Water Splitting and Photocatalytic Dye Removal Activities of Camphor Sulfonic Acid Doped Polyaniline -WO3-MWCNT Ternary Nanocomposite." *Journal of Materials Science & Technology*. https://doi.org/10.1016/j.jmst.2019.08.020.

Huang, Ruihua, Bingchao Yang, Dongsheng Zheng, Guohua Chen, and Congjie Gao. 2012. "Treatment of Textile Dye Effluent Using a Self-Made Positively Charged Nanofiltration Membrane." *Journal Wuhan University of Technology, Materials Science Edition* 27 (2): 199–202. https://doi.org/10.1007/s11595-012-0436-0.

Hunger, Klaus, ed. 2002. *Industrial Dyes. Industrial Dyes*. Wiley. https://doi.org/10.1002/3527602011.

Iram, Mahmood, Chen Guo, Yueping Guan, Ahmad Ishfaq, and Huizhou Liu. 2010. "Adsorption and Magnetic Removal of Neutral Red Dye from Aqueous Solution Using Fe3O4 Hollow Nanospheres." *Journal of Hazardous Materials* 181 (1–3): 1039–50. https://doi.org/10.1016/j.jhazmat.2010.05.119.

J. Fendi, Wedad, and Juman A. Naser. 2018. "Adsorption Isotherms Study of Methylene Blue Dye on Membranes from Electrospun Nanofibers." *Oriental Journal of Chemistry* 34 (6): 2884–94. https://doi.org/10.13005/ojc/340628.

Kannan, Nagarethinam, and Mariappan Meenakshi Sundaram. 2001. "Kinetics and Mechanism of Removal of Methylene Blue by Adsorption on Various Carbons - A Comparative Study." *Dyes and Pigments* 51 (1): 25–40. https://doi.org/10.1016/S0143-7208(01) 00056-0.

Khajeh, Mostafa, Sophie Laurent, and Kamran Dastafkan. 2013. "Nanoadsorbents: Classification, Preparation, and Applications (with Emphasis on Aqueous Media)." *Chemical Reviews* 113 (10): 7728–68. https://doi.org/10.1021/cr400086v.

Khosravi, Alireza, Mohammad Karimi, Hadiyeh Ebrahimi, and Narges Fallah. 2020. "Sequencing Batch Reactor/Nanofiltration Hybrid Method for Water Recovery from Textile Wastewater Contained Phthalocyanine Dye and Anionic Surfactant." *Journal of Environmental Chemical Engineering* 8 (2): 103701. https://doi.org/10. 1016/j.jece.2020.103701.

Kim, Ye-sol, and Jin-hyun Kim. 2018. "Isotherm, Kinetic and Thermodynamic Studies on the Adsorption of Paclitaxel onto Sylopute." *The Journal of Chemical Thermodynamics*, no. October. https://doi.org/10.1016/j.jct.2018.10.005.

Kiwaan, H. A., T. M. Atwee, E. A. Azab, and A. A. El-Bindary. 2020. "Photocatalytic Degradation of Organic Dyes in the Presence of Nanostructured Titanium Dioxide." *Journal of Molecular Structure* 1200 (January): 127115. https://doi.org/10.1016/j.molstruc.2019. 127115.

Lam, Y. L., C. W. Kan, and C. W. M. Yuen. 2012. "Application of Catalyst in Textile Wet Processes." *Research Journal of Textile and Apparel* 16 (1): 10–23. https://doi.org/10.1108/RJTA-16-01-2012-B002.

Long, Zhihang, Yingqing Zhan, Fei Li, Xinyi Wan, Yi He, Chunyan Hou, and Hai Hu. 2017. "Hydrothermal Synthesis of Graphene Oxide/Multiwalled Carbon Nanotube/Fe3O4 Ternary Nanocomposite for Removal of Cu (II) and Methylene Blue." *Journal of Nanoparticle Research* 19 (9): 318. https://doi.org/10.1007/s11051-017-4014-4.

Madrakian, Tayyebeh, Abbas Afkhami, Mazaher Ahmadi, and Hasan Bagheri. 2011. "Removal of Some Cationic Dyes from Aqueous Solutions Using Magnetic-Modified Multi-Walled Carbon Nanotubes."

Journal of Hazardous Materials 196 (November): 109–14. https://doi.org/10.1016/j.jhazmat.2011.08.078.

Mahmoud, Hala R., Sahar A. El-Molla, and M. Saif. 2013. "Improvement of Physicochemical Properties of Fe2O3/MgO Nanomaterials by Hydrothermal Treatment for Dye Removal from Industrial Wastewater." *Powder Technology* 249: 225–33. https://doi.org/10.1016/j.powtec.2013.08.021.

Min, Ohm Mar, Li Ngee Ho, Soon An Ong, and Yee Shian Wong. 2015. "Comparison between the Photocatalytic Degradation of Single and Binary Azo Dyes in TiO2 Suspensions under Solar Light Irradiation." *Journal of Water Reuse and Desalination* 5 (4): 579–91. https://doi.org/10.2166/wrd.2015.022.

Mondal, Mrinmoy, and Sirshendu De. 2016. "Treatment of Textile Plant Effluent by Hollow Fiber Nanofiltration Membrane and Multi-Component Steady State Modeling." *Chemical Engineering Journal* 285: 304–18. https://doi.org/10.1016/j.cej.2015.10.005.

Monsef Khoshhesab, Zahra, and Samira Souhani. 2018. "Adsorptive Removal of Reactive Dyes from Aqueous Solutions Using Zinc Oxide Nanoparticles." *Journal of the Chinese Chemical Society* 65 (12): 1482–90. https://doi.org/10.1002/jccs.201700477.

Mortazavian, Soroosh, Ali Saber, and David E. James. 2019. "Optimization of Photocatalytic Degradation of Acid Blue 113 and Acid Red 88 Textile Dyes in a Uv-c/Tio 2 Suspension System: Application of Response Surface Methodology (Rsm)." *Catalysts* 9 (4): 360. https://doi.org/10.3390/catal9040360.

Moussavi, Gholamreza, and Maryam Mahmoudi. 2009. "Removal of Azo and Anthraquinone Reactive Dyes from Industrial Wastewaters Using MgO Nanoparticles." *Journal of Hazardous Materials* 168 (2–3): 806–12. https://doi.org/10.1016/j.jhazmat.2009.02.097.

Nasreen, Shaik Anwar Ahamed Nabeela, Subramanian Sundarrajan, Syed Abdulrahim Syed Nizar, Ramalingam Balamurugan, and Seeram Ramakrishna. 2013. "Advancement in Electrospun Nanofibrous Membranes Modification and Their Application in Water Treatment."

Membranes 3 (4): 266–84. https://doi.org/10.3390/membranes 3040266.

Ngomsik, Audrey Flore, Agnès Bee, Micheline Draye, Gérard Cote, and Valérie Cabuil. 2005. "Magnetic Nano- and Microparticles for Metal Removal and Environmental Applications: A Review." *Comptes Rendus Chimie* 8 (6-7 SPEC. ISS.): 963–70. https://doi.org/10. 1016/j.crci.2005.01.001.

Nirumand, Ladan, Saeed Farhadi, Abedin Zabardasti, and Alireza Khataee. 2018. "Synthesis and Sonocatalytic Performance of a Ternary Magnetic MIL-101(Cr)/RGO/ZnFe2O4 Nanocomposite for Degradation of Dye Pollutants." *Ultrasonics Sonochemistry* 42 (October 2017): 647–58. https://doi.org/10.1016/j.ultsonch.2017. 12.033.

Ong, Yee Kang, Fu Yun Li, Shi Peng Sun, Bai Wang Zhao, Can Zeng Liang, and Tai Shung Chung. 2014. "Nanofiltration Hollow Fiber Membranes for Textile Wastewater Treatment: Lab-Scale and Pilot-Scale Studies." *Chemical Engineering Science* 114 (July): 51–57. https://doi.org/10.1016/j.ces.2014.04.007.

Pal, Umapada, Alberto Sandoval, Sergio Isaac Uribe Madrid, Grisel Corro, Vivek Sharma, and Paritosh Mohanty. 2016. "Mixed Titanium, Silicon, and Aluminum Oxide Nanostructures as Novel Adsorbent for Removal of Rhodamine 6G and Methylene Blue as Cationic Dyes from Aqueous Solution." *Chemosphere* 163: 142–52. https://doi.org/ 10.1016/j.chemosphere.2016.08.020.

Panda, Swapna Rekha, and Sirshendu De. 2015. "Performance Evaluation of Two Stage Nanofiltration for Treatment of Textile Effluent Containing Reactive Dyes." *Journal of Environmental Chemical Engineering* 3 (3): 1678–90. https://doi.org/10.1016/j.jece.2015. 06.004.

Patra, Santanu, Ekta Roy, Rashmi Madhuri, and Prashant K. Sharma. 2016. "Agar Based Bimetallic Nanoparticles as High-Performance Renewable Adsorbent for Removal and Degradation of Cationic Organic Dyes." *Journal of Industrial and Engineering Chemistry* 33: 226–38. https://doi.org/10.1016/j.jiec.2015.10.008.

Potle, Vinit D., Sachin R. Shirsath, Bharat A. Bhanvase, and Virendra Kumar Saharan. 2020. "Sonochemical Preparation of Ternary RGO-ZnO-TiO2 Nanocomposite Photocatalyst for Efficient Degradation of Crystal Violet Dye." *Optik* 208 (December 2019). https://doi.org/10.1016/j.ijleo.2020.164555.

Qu, Xiaolei, Pedro J.J. Alvarez, and Qilin Li. 2013. "Applications of Nanotechnology in Water and Wastewater Treatment." *Water Research* 47 (12): 3931–46. https://doi.org/10.1016/j.watres. 2012.09.058.

Raghavan, Nivea, Sakthivel Thangavel, and Gunasekaran Venugopal. 2015. "Enhanced Photocatalytic Degradation of Methylene Blue by Reduced Graphene-Oxide/Titanium Dioxide/Zinc Oxide Ternary Nanocomposites." *Materials Science in Semiconductor Processing* 30: 321–29. https://doi.org/10.1016/j.mssp.2014.09.019.

Reza, Khan Mamun, ASW Kurny, and Fahmida Gulshan. 2017. "Parameters Affecting the Photocatalytic Degradation of Dyes Using TiO2: A Review." *Applied Water Science* 7 (4): 1569–78. https://doi.org/10.1007/s13201-015-0367-y.

Riaz, Qamar, Madiha Ahmed, Muhammad Nadeem Zafar, Muhammad Zubair, Muhammad Faizan Nazar, Sajjad Hussain Sumrra, Iqbal Ahmad, and Ahmad Hosseini-Bandegharaeic. 2020. "NiO Nanoparticles for Enhanced Removal of Methyl Orange: Equilibrium, Kinetics, Thermodynamic and Desorption Studies." *International Journal of Environmental Analytical Chemistry*, January, 1–20. https://doi.org/10.1080/03067319.2020.1715383.

Sá, Arsénio de, Ana S. Abreu, Isabel Moura, and Ana Vera Machado. 2017. *Polymeric Materials for Metal Sorption from Hydric Resources. Water Purification*. Elsevier Inc. https://doi.org/10.1016/b978-0-12-804300-4.00008-3.

Sadegh, Hamidreza, Gomaa A. M. Ali, Vinod Kumar Gupta, Abdel Salam Hamdy Makhlouf, Ramin Shahryari-ghoshekandi, Mallikarjuna N. Nadagouda, Mika Sillanpää, and Elżbieta Megiel. 2017. "The Role of Nanomaterials as Effective Adsorbents and Their Applications in Wastewater Treatment." *Journal of Nanostructure in Chemistry* 7 (1): 1–14. https://doi.org/10.1007/s40097-017-0219-4.

Sadeghi-Kiakhani, Mousa, Mokhtar Arami, and Kamaladin Gharanjig. 2013. "Dye Removal from Colored-Textile Wastewater Using Chitosan-PPI Dendrimer Hybrid as a Biopolymer: Optimization, Kinetic, and Isotherm Studies." *Journal of Applied Polymer Science* 127 (4): 2607–19. https://doi.org/10.1002/app.37615.

Saeed, Khalid, Noor Zada, and Idrees Khan. 2019. "Photocatalytic Degradation of Alizarin Red Dye in Aqueous Medium Using Carbon Nanotubes/Cu–Ti Oxide Composites." *Separation Science and Technology (Philadelphia)* 54 (16): 2729–37. https://doi.org/10.1080/01496395.2018.1552296.

Savage, Nora, and Mamadou S. Diallo. 2005. "Nanomaterials and Water Purification: Opportunities and Challenges." *Journal of Nanoparticle Research* 7 (4–5): 331–42. https://doi.org/10.1007/s11051-005-7523-5.

Stoyanova, Angelina, Albena Bachvarova-Nedelcheva, and Reni Iordanova. 2018. "Photocatalytic Degradation of Two Azo-Dyes in Single and Binary Mixture by La Modified Tio2." *Journal of Chemical Technology and Metallurgy* 53 (6): 1173–78.

Suja, P. S., C. R. Reshmi, P. Sagitha, and A. Sujith. 2017. "Electrospun Nanofibrous Membranes for Water Purification." *Polymer Reviews* 57 (3): 467–504. https://doi.org/10.1080/15583724.2017.1309664.

Tandon, Praveen K. 2014. "Catalysis : A Brief Review on Nano-Catalyst." *Journal of Energy and Chemical Engineering* 2 (3): 106–15.

Tavangar, Tohid, Kamran Jalali, Mohammad Amin Alaei Shahmirzadi, and Mohammad Karimi. 2019. "Toward Real Textile Wastewater Treatment: Membrane Fouling Control and Effective Fractionation of Dyes/Inorganic Salts Using a Hybrid Electrocoagulation – Nanofiltration Process." *Separation and Purification Technology* 216 (January): 115–25. https://doi.org/10.1016/j.seppur.2019.01.070.

Venkatesha, T. G., R. Viswanatha, Y. Arthoba Nayaka, and B. K. Chethana. 2012. "Kinetics and Thermodynamics of Reactive and Vat Dyes Adsorption on MgO Nanoparticles." *Chemical Engineering Journal* 198–199: 1–10. https://doi.org/10.1016/j.cej.2012.05.071.

Wang, Jung Pin, Hsi Chi Yang, and Chien Te Hsieh. 2011. "Adsorption of Phenol and Basic Dye on Carbon Nanotubes/Carbon Fabric Composites from Aqueous Solution." *Separation Science and Technology* 46 (2): 340–48. https://doi.org/10.1080/01496395.2010. 508066.

Xu, Yuanting, Jianxu Bao, Xu Zhang, Wusha Li, Yi Xie, Shudong Sun, Weifeng Zhao, and Changsheng Zhao. 2019. "Functionalized Polyethersulfone Nanofibrous Membranes with Ultra-High Adsorption Capacity for Organic Dyes by One-Step Electrospinning." *Journal of Colloid and Interface Science* 533 (January): 526–38. https://doi.org/10.1016/j.jcis.2018.08.072.

Yu, Fei, Junhong Chen, Lu Chen, Jing Huai, Wenyi Gong, Zhiwen Yuan, Jinhe Wang, and Jie Ma. 2012. "Magnetic Carbon Nanotubes Synthesis by Fenton's Reagent Method and Their Potential Application for Removal of Azo Dye from Aqueous Solution." *Journal of Colloid and Interface Science* 378 (1): 175–83. https://doi.org/10.1016/j.jcis.2012.04.024.

Zafar, Muhammad Nadeem, Qamar Dar, Faisal Nawaz, Muhammad Naveed Zafar, Munawar Iqbal, and Muhammad Faizan Nazar. 2019. "Effective Adsorptive Removal of Azo Dyes over Spherical ZnO Nanoparticles." *Journal of Materials Research and Technology* 8 (1): 713–25. https://doi.org/10.1016/j.jmrt.2018.06.002.

Zhang, Wei, Linda Zou, and Lianzhou Wang. 2010. "Visible-Light Assisted Methylene Blue (MB) Removal by Novel TiO2/Adsorbent Nanocomposites." *Water Science and Technology* 61 (11): 2863–71. https://doi.org/10.2166/wst.2010.196.

Zhao, Shuang, Peng Song, Zhan Wang, and Hongtai Zhu. 2018. "The PEGylation of Plant Polyphenols/Polypeptide-Mediated Loose Nanofiltration Membrane for Textile Wastewater Treatment and Antibacterial Application." *Journal of the Taiwan Institute of Chemical Engineers* 82: 42–55. https://doi.org/10.1016/j.jtice.2017. 11.005.

Zhao, Xin, Lu Lv, Bingcai Pan, Weiming Zhang, Shujuan Zhang, and Quanxing Zhang. 2011. "Polymer-Supported Nanocomposites for Environmental Application: A Review." *Chemical Engineering Journal* 170 (2–3): 381–94. https://doi.org/10.1016/j.cej.2011.02.071.

Zheng, Yinping, Sanchuan Yu, Shi Shuai, Qing Zhou, and Qibo Cheng. 2013. "Color Removal and COD Reduction of Biologically Treated

Textile Ef Fl Uent through Submerged Fi Ltration Using Hollow Fi Ber Nano Fi Ltration Membrane." *Desalination* 314: 89–95. https://doi.org/10.1016/j.desal.2013.01.004.

Zheng, Yinping, Sanchuan Yu, Shi Shuai, Qing Zhou, Qibo Cheng, Meihong Liu, and Congjie Gao. 2013. "Color Removal and COD Reduction of Biologically Treated Textile Effluent through Submerged Filtration Using Hollow Fiber Nanofiltration Membrane." *Desalination* 314 (April): 89–95. https://doi.org/10.1016/j.desal.2013.01.004.

Zhu, H. Y., R. Jiang, L. Xiao, and G. M. Zeng. 2010. "Preparation, Characterization, Adsorption Kinetics and Thermodynamics of Novel Magnetic Chitosan Enwrapping Nanosized γ-Fe2O3 and Multi-Walled Carbon Nanotubes with Enhanced Adsorption Properties for Methyl Orange." *Bioresource Technology* 101 (14): 5063–69. https://doi.org/10.1016/j.biortech.2010.01.107.

In: Challenges and Opportunities … ISBN: 978-1-53618-770-0
Editor: Wallace G. Tarrant

Chapter 2

THE USE OF NANOMATERIALS IN FUNCTIONAL TEXTILES

***Mahmut Tas*[1] *and Abdurrahman Telli*[2,*]**
[1]Faculty of Engineering,
The University of Nottingham, Nottingham, UK
[2]Department of Textile Engineering,
Cukurova University, Adana, Turkey

ABSTRACT

Nanosized materials are becoming more and more popular in almost every field of the industry. Technical textiles are one of the major applications of nanomaterials and there are many commercialized nanoparticles functionalized textiles on the market such as electromagnetic interference shielding materials, flame retardant textiles, UV-protective fabrics, etc. The compatibility of conventional textile processes for nanoparticle applications and reduced nanomaterial production costs are made much faster to produce commercialized nanoparticle functionalized textiles. Nanoclays, carbon-based nanomaterials, and metal-based nanopowders are the most commonly used nanomaterials in functional

* Corresponding Author's Email: atelli@cu.edu.tr.

textiles. It is possible to produce nanomaterial functionalized textile by adding nanoparticles into polymer melt or solution for synthetic fibres, or surface treatment of fiber or textile materials. In this chapter of the book, the benefits of using nanomaterials for functionalized textile production will be presented. Applications of nanomaterials in the most known functional technical textiles, (flame retardant, antimicrobial, electromagnetic interference shielding, UV-protective, superhydrophobic, and wrinkle resistive) will be discussed.

Keywords: functional textiles, applications of nanoparticles, flame retardant textiles, medical textiles, protective textiles

INTRODUCTION

Nanomaterials are defined as the materials with at least one of the dimensions smaller than 100 nm [1, 2]. These materials have unique properties such as high surface area, good mechanical properties, higher reactivity, etc. The industries of renewable energy [3], electronics [4], biomedical [5], food agriculture [6], and textile [7] are comprising some main usage areas of nanomaterials. Although there are many commercialized nanomaterial-functionalized textile products, the expected level of nanomaterial usage has not been reached yet, and the researchers are working to find out new functional textiles using nanomaterials. It is expected to achieve around $15 billion of the market share of the nanomaterial-functionalized textiles by 2024 which would likely offer a huge opportunity to the textile industry. It is possible to impart functional properties such as flame retardancy [8] antibacterial activity [9], UV-protection [10], electromagnetic shielding efficiency [11], electrical conductivity [12], and many more beneficial properties to conventional textile material. A variety of nanoparticles such as polymeric materials, carbon-based structures, and metal salts are used to provide functional properties, with large scale production methods for instance padding, melt mixing, and chemical treatments. Some of the main advantages of using nanomaterials to functionalize the textile structures are fewer requirements for materials and chemicals, easy application processes, and the possibility to keep the properties of the textile materials such as breathability. This study involves a summary of the use of nanomaterials in functional textiles. The advantages and disadvantages of using nanomaterials in functional textiles are discussed in detail.

Antimicrobial Textile Applications of Nanomaterials

Antimicrobial textiles are specifically designed materials that can inhibit or slows down the growth of several types of microorganisms such as bacteria, viruses, fungi, etc. They have a wide range of usage areas such as medical applications (clinical, first aid, hygiene, and surgical purposes), protective clothing, home textiles, and items of clothing (sportswear, underwear, socks, etc.). When a microorganism grows on the textile surface, undesirable effects occur such as unpleasant odor, reduced mechanical properties, and stains and it can easily contaminate the user. Antimicrobial textiles can prevent this as well as slow down the spread of infection.

Nanotechnology made it possible to produce comfortable antimicrobial textiles with high durability (depend on the physical or chemical adhesion between the nanomaterial and textile surface) thanks to high surface energy and large surface area of nanomaterials. The most important methods to produce antimicrobial textile materials can be classified as follows [13],

- Solution or melt mixing of polymer and antimicrobial nanomaterials
- Chemical grafting of nanomaterials onto the textile surface
- Padding-dry cure or exhaustion procedures
- Foaming or spraying techniques
- Surface activation and chemical bonding of nanomaterials to the textile surface
- Sol-gel or microencapsulation approaches

To prove the antimicrobial effect, there are several standardized methods and different requirements of experimental setups. For the determination of antibacterial effect following standard test methods are the most commonly used ones [14],

- AATC 100 (quantitative test method)
- ASTM E2149 (activity of immobilized antimicrobial agents under dynamic contact conditions)
- AATCC 147 (parallel streak methods)
- DIN EN ISO 20743 (absorption, transfer or printing methods)

Additionally, for antifungal effect determination of textiles, AATCC 30 and DIN EN 14119 standards are known as the most popular methods.

Several types of nanomaterials can be used to impart antimicrobial properties to textile materials. However, silver nanoparticles are considered as the most effective materials which have high potential to kill or inhibit more than 650 types of bacteria, microorganisms, and viruses [15]. It was also used for wound and burn healing applications for centuries before the discovery of penicillin in the form of silver salts and metallic silver [16]. Furthermore, silver has low toxicity for the human body and has good biocompatibility. Silver nanoparticles can be applied by embedding them inside the fibres during the extrusion process for man-made fibres, or surface modification methods are applied using different techniques including padding, in situ synthesis of silver nanoparticles, spraying etc.

Although the antimicrobial effect mechanism of silver nanoparticles is not understood clearly, it is believed that the released silver ions from silver nanoparticles, when exposed to the oxygen or water, has a great contribution. Equation (1) shows the ion release of silver nanoparticles with the help of oxygen and water [15],

$$O_{2(aq)} + 4H_3O^+ + 4Ag_{(s)} \rightarrow 4Ag^+_{(aq)} + 6H_2O \qquad \text{Eq. (1)}$$

Silver nanoparticles are used almost for every type of textiles specifically cotton [17-20], wool [21, 22], polyester [23], polypropylene

[24], and polyamide [25] in the form of fibre, yarn, and fabric. Some of the studies report that the breathability of the fabrics is not affected by the application of silver nanoparticles and also, the size of the silver nanoparticles has an important effect on the antimicrobial effect.

Copper and some of its compounds (copper (I) oxide, copper (II) oxide, and copper complexes) has also different levels of antimicrobial effect and are used to impart antimicrobial effect to conventional textiles, specifically cellulose-based fabrics such as cotton, tencel and viscose [26]. They are synthesized easily from copper precursors using two-steps chemical reaction, reduction, and oxidation. Compared to silver nanoparticles, the synthesis of copper nanoparticles cost much less because the required precursors are inexpensive. The same application methods with silver nanoparticles are used to apply copper nanoparticles to the textile surface. Additionally copper considered safe for the human body due to its non-toxic nature.

Due to their photocatalytic effect, TiO_2 and ZnO are used as an antimicrobial agent for textiles [10, 27]. They are also provided UV protective and self-cleaning properties when used with textile materials. Thus, it is possible to achieve multifunctionality using TiO_2 and ZnO nanoparticles. Both materials are effective, especially against bacteria.

Chitosan, a deacetylated form of chitin, is one of the most abundant polymers in nature. The chemical structure of chitosan is given in Figure 1. They are produced from shrimp shells, crab shells, and some types of fungi [28]. They considered promising antimicrobial material due to their high potential to inhibit and kills various microorganisms. It is possible to produce chitosan nanoparticles using ionic gelation, microemulsion, reverse micellar, and complex coacervation methods. Several studies reported the antibacterial effect of chitosan nanoparticles treated textiles. Besides, to increase the antibacterial effect, hybrid structures such as silver\chitosan [29] or ZnO\Chitosan [30] are used and an improved antibacterial effect is achieved.

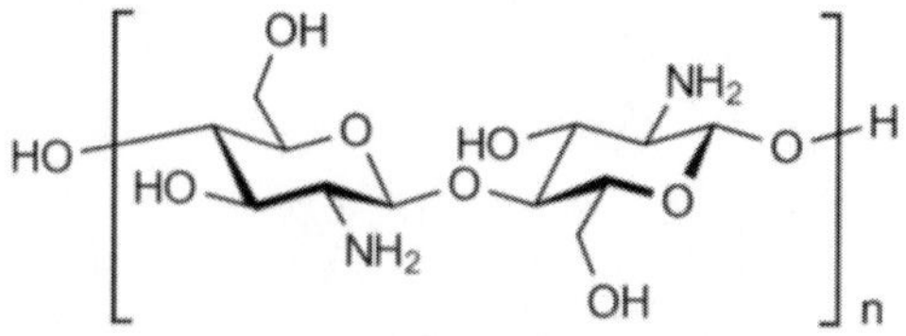

Figure 1. Chemical structure of Chitosan.

Although antibacterial textiles comprise a major part of nanomaterials treated technical textiles, there are still lots of gaps in the literature. Affordable, environmentally friendly, biocompatible, and durable antimicrobial textiles still under investigation by the researchers.

FLAME RETARDANT TEXTILE APPLICATIONS OF NANOMATERIALS

Most of the man-made and natural fibers used in the textile industry are known as flammable materials. However, low flammability is a critical property for some applications such as protective clothing, home textile products (carpets, curtains, mattresses, etc.) and industrial textiles (Figure 2) [31].

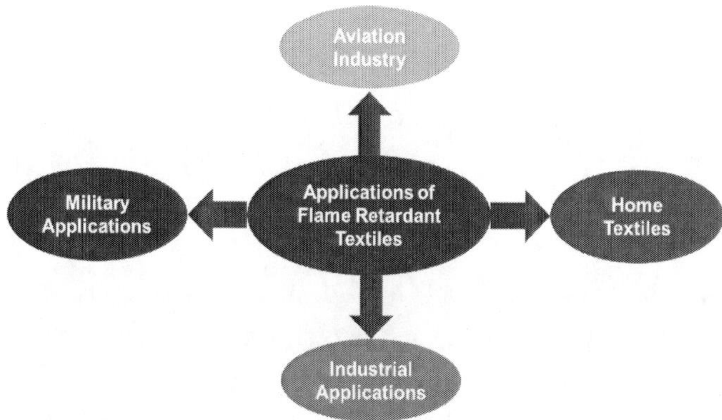

Figure 2. Main applications of flame retardant textiles.

Table 1. LOI values of some common textile materials

Material Type	LOI Values
Cotton, Viscose, and Linen	17-20
Wool	24-27
Silk	23
Polyester	22
Polyethylene	17.5
Polypropylene	18-20
Polyamide	20
Aramide	25-28
Teflon	95

Flammability can be defined based on flame spread rate, the release of heat, and ignialibility [32]. To determine these behaviors, the tests of thermogravimetric analysis, UL94 vertical burning test, cone calorimetry, and limiting oxygen index (LOI) are the most commonly used techniques [33]. The LOI values of some textile materials are given in Table 1.

There are three main approaches to reduce the flammability of textile materials (i) use of polymer/flame retarder composite fibres, (ii) modification of the textile using graft polymerization, and (iii) surface treatment methods. Most of the commercial flame retardant textile systems are consisting of halogen-included compounds which are environmentally hazardous. Alternatively, using flame retardant nanoparticles is a continuously growing trend to reduce the flammability of textile materials. Therefore, there are many reports published in the literature about the nanoparticle-assisted flame retardant textile materials.

The application technique, the type, geometry, and amount of the nanoparticles used are the most important parameters which determine the level of flammability. Melt or solution mixing, padding, and sonochemical coating are the most practice and used methods to prepare nanoparticle-assisted flame retardant textiles.

Nanoclays are the most popular type of flame retardant nanoparticles and known as environmentally friendly and cost effective [34]. There are many reported studies in the literature with different approaches to incorporate nanoclays with textile materials to reduce the flammability [35-

37]. Although Montmorillonite is the most commonly used one, there are several other classes of nanoclays such as sepiolite, kaolinite, bentonite, and halloysite. Nanoclays are generally used in a polymer in a polymer matrix therefore, dispersibility of nanoclays in the polymer matrix is a critical parameter that affects the efficiency of reduction of flammability. To improve the dispersibility of nanoclays in the polymer matrix, surface modification steps are carried out with organic cationic surfactants which provide better wettability.

The mechanism beyond the flame retardancy of nanoclay/Polymer nanocomposites has been discussed Yang et al., Different compositions of Polypropylene/clay nanocomposites prepared using a twin-screw extruder and thermal and combustion properties investigated. It was found that the improvement inflammability was mostly about the condensed-phase flame retardant mechanism and the decrease of heat release rate is related to delay of thermal-oxidative decomposition [38].

Carbon-based materials are another promising material class for the flame retardant textiles [33]. Fullerenes [39], carbon nanotubes, and chemically modified forms [40], graphene [41], and carbon black reported as flame retarder nanomaterials in the literature. However there is a similar problem with nanoclays, carbon-based materials are not miscible with polymer materials and to prepare homogenous mixtures, a surface modification process should be carried out.

Generally, these materials can provide not only flame retardancy but also contribute to other important properties such as electrical conductivity, mechanical properties, thermal conductivity, etc. Thus, they are quite useful to design multifunctional textile structures.

The main mechanisms of flame retardancy effect of carbon-based materials can be summarized under two explanations. 1) they can contribute to the formation of a protective char layer without defects and cracks that can shield the textile materials and 2) absorption of highly active radicals by some of the carbon-based materials which slow down the burning process.

Unlike the carbon black and graphite, due to the high cost of carbon nanotubes, graphene and fullerenes, there is no commercial application on flame retardant textiles yet. However, large scale and cost-effective production methods can make it possible to use these kinds of materials for flame retardant applications.

Polyhedral Oligomeric Silsesquioxane (POSS) [42] nanoparticles are another type of flame retardant that attracts attention especially because of their excellent thermo oxidative stability, environmental compatibility, and heat resistance properties and have the general formula of $(RSiO_{1.5})_n$. They have an average size of 1-3 nm and in addition, it is possible to chemically modify the surface of POSS thanks to organic R groups, which can be designed as reactive or non-reactive, that surrounds the nanoparticle.

Although POSS itself has great potential for flame retardant applications, it costs much higher than most of the commercial flame retardants and as a result, it can be used only for particular applications which cost is not important such as aviation, military, etc. However, POSS also has a great synergistic effect when used with other types of flame retardant materials even for very low additive amounts. The mechanism of POSS can be summarized as the formation of superficial char layers. During burning, POSS transforms to thermally and oxidatively stable ceramic surfaces and shields the material from the heat.

Metal-based nanomaterials such as TiO_2 [43, 44], ZnO [45, 46], Fe_2O_3 [47] BPO_4 [48], Al_2O_3 [49] are also reported as flame retardant materials for textile polymers. There are several suggested mechanisms of metal-based nanomaterials[32]: they can (i) limit the mobility of polymer molecules, (ii) absorb the free radicals (iii) shield the heat and mass transfer, and (iv) reduce the heat release.

UV-Protective Textile Applications of Nanomaterials

Light is an essential factor for all living creatures on our planet. Sun energy radiates at different wavelengths. Sunlight that consists of 5% UV, 50% visible, and 45% infrared radiation reach the earth by passing the ozone layer. There are three types of UV rays classified by wavelengths and measured in "Nanometer, nm": UV-A UV-B and UV-C. Exposure of people to short-wavelength sunlight (intense UV light) can cause several problems. UV-A has a long wavelength (320-400 nm) and always reaches the earth. It makes up about 95% of the UV radiation reaching the earth's surface. It helps produce vitamin D for living creatures. As a result of prolonged exposure, UV-A rays in the long-wavelength penetrate deep into the skin. It causes aging of the skin, loss of elasticity, and wrinkles. It is known to increase skin cancer development. UV-B (290-320 nm) does not have a wavelength as long as UV-A. Since it has a shorter wavelength, UV-B damages human and plant cells more than UV-A. The entire of UV-B is not reached to earth because some are filtered by the ozone layer. UV-B rays can penetrate less of the skin in mm but cause permanent pigmentation here, in the epidermis. It can also damage our cells and DNA. So it can cause skin cancer and cataracts in our eyes. UV-C has the shortest wavelength (190-290 nm). It is the most dangerous, but they are all absorbed by the ozone layer [50-52].

Apart from the natural photoprotection mechanisms of humans, the most important assistant in UV protection is textile products. Textile products reflect some of the UV rays, absorb some of them, and pass some of them. The value of SPF (Sun Protection Factor) or UPF (Ultraviolet Protection Factor), which expresses how long it can be protected by a textile until the first redness is detected in skin, can be determined according to in vivo or in vitro methods. The time spent in minutes for the first redness seen on the skin exposed to the sun at sunny noon is the starting point of the UPF. It may show variation according to the body structure. For example, it is understood from UPF that a person who has the skin with the first redness after 7 minutes can stay under the sun up to (7x10) for 70 minutes using a garment with UPF10. Fine summer knitwear usually has a weak ultraviolet protection factor of 5-15 because ultraviolet rays can pass directly through the holes in the loosely knitted. It is possible to increase the UPF value by changing the fabric structure and content. Although the UPF value can be increased up to 30 with a dark and tight structure or a multi-layered textile product, nobody will want to use such a product as this will increase body temperature. For this reason, the development of UV protective properties of textiles is one of the major research topics [52, 53].

Besides, UV rays cause color fading of textile materials and loss of strength. The UV photo-degradation problem mostly applies to outdoor technical textiles, daily wear, automobile interiors, and indoor textiles near windows (such as upholstery, curtains, and blinds). Outdoor living spaces are becoming more and more important in homes and commercial areas. In recent years, especially, the commercial awning industry is reborn thanks to awnings used in new shopping malls and store chain designs.

PAC fibers are the most resistant to light and outdoor conditions among all-important natural and artificial fibers used in textiles. 95% of cotton, 38% of polyester, and 18% of PAC are lost while the breaking strength of silk, wool, regenerated cellulose, polyamide fibers, which are left in the open air for a year, decreases to zero. Polyester fibers have good resistance to outdoor weather conditions. Since polyester fibers have higher initial strength (such as rupture and rubbing) compared to PAC, the use of polyester fibers plays an important role in the construction of materials such as awnings, tarpaulins, sails, and ropes. Furthermore, the curtains produced from polyester fibers are slightly higher resistant to sun rays since the rays acting primarily on polyester fibers are UV rays with a wavelength of 300-330 nm and a significant part of them are absorbed by the window glass. The resistance of dull polyester fibers to light and outdoor conditions is lower compared to bright polyester types without pigment. However, the situation is very different in other natural and artificial fibers. For example, rays (UV) with a wavelength shorter than 380nm cause yellowing of wool fibers. These effects become more apparent depending on oxygen and water. The reason why the wool fibers turn yellow under the light effect is related to the formation of colored decomposition products by breaking down some amino acid building blocks. In wool fibers bleached with oxidizing agents (e.g., H2O2), the risk of yellowing increases, even more, when wet. In wool fibers bleached with reducing agents (e.g., hydrogen sulfide), a similar increase is not seen. The tensile and friction resistance of woolen products, which are exposed to sunlight for a long time, is significantly reduced. The tensile strength of specimens with sunlight up to 900 hours decreases by 70%. In the analyzes after irradiation, it has been observed that there is a big increase in the number of amine groups in the fibers. Accordingly, there are breaks in the macro-peptide chains with the effect of light and temperature. However, these breaks do not occur in peptide bonds, they occur in covalent bonds in the α-state. Polyamide fibers, which are kept in sunlight for a long time, also show a great decrease in their tensile strength and white fibers turn yellow over time. The photo-oxidative damage of polyamide fibers leads to a hidden yellowing. The yellowing is not immediately apparent, it is evident after a long time or short heating. The number of carboxyl groups in photo-oxidative damaged fibers increases, and the number of amino groups decreases. Adipic acid, pyrrole, and formic acid groups are detected at the ends of the oligomers formed in the damaged fibers. This indicates that the damage started with the radicals formed in the -CH2 group adjacent to the -NH group, and eventually there were also breaks in the -NH-CO- group [54].

UV absorbers are used to increase the ultraviolet protection factor (UPF) in fabrics. These products have a high affinity for UV waves and turn attacks by UV into harmless heat. UV absorbers can be applied to the fabric by coating. The second option is the application of the finishing processing of fabrics with UV absorbers. The third option is the dosing of the additive containing UV absorber to polymers by using dyeing technology in the melt spinning method. UV absorbers that can be applied to protect against UV radiation are two types, inorganic and organic. Inorganic UV absorbers are preferred over organic UV absorbers because they are non-toxic and chemically stable under the influence of high temperature and UV. Inorganic UV absorbers are often certain semiconductor oxides such as TiO_2, ZnO, SiO_2, Al_2O_3. Among them, TiO_2 and ZnO are widely used [55].

TiO_2 is used for brightness between 1-3% in polyester fibers during fiber production. UV absorption ability of polyester increases when this rate is increased to 3-5%. Polyamide fibers dulled with TiO2 are much faster and more damaged from light according to polyester. That is, titanium dioxide plays the role of a photo-sensibilization agent. To reduce the damage caused by the polyamide fibers from the light, a certain amount of stabilizer is added to the commercially available fibers. Mn^{2+}, Cu^{1+}, Cr^{3+} compounds, or organic compounds such as phenol, aromatic amines ... are used as stabilizers. These are estimated to reduce damage by converting hydro-peroxides formed by the action of light and oxygen into the ionic state. Some dispersion dyes are also known to exhibit a weak stabilizing effect [5]. In coating and finishing applications, preferred UV absorbers for cotton and cotton blends are oxaldianilides containing two water-soluble reactive groups. It is applied to the fabric as 15-50 g/lt in the pad-batch method. The ratio is 1-4% in the jet or over-flow processing with reactive or direct dyestuffs. It has been found that it is very good both intake by fiber and fixing properties. Mono-sulfonated benzotriazole derivative UV absorbers are preferred for wool, silk, polyamide, and their blends. These are used in the same process conditions at the rate of 1-3% during dyeing with acid dyes. For polyester and its blends, 1-3% benzotriazole derivative UV absorbers are used during the dyeing with disperse dyes [50, 56].

However, they can show photocatalytic effect and cause deterioration in the integrity of polymeric materials used in fabrics and coatings while UV absorbers convert UV rays into heat. Also, washing fastness is of great importance in both natural and synthetic fabric applications. In most commercially or theoretically recommended applications, the level of UV protection after washing is significantly reduced. At this point, nano UV absorbers come to the forefront both in reducing or eliminating photocatalytic capacity and in improving washing fastness. Among the nanoparticles, especially clay nanoparticles and metal oxide nanoparticles find use in textiles. Metal oxides such as TiO_2, Al_2O_3, ZnO, and MgO can be applied to textiles during polymer fiber drawing, by electrostatic methods or by spray coating. The most efficient method is to prepare polymer granules and nanoparticles as masterbatches with the help of twin-screw extruders. The prepared masterbatches are added during the dyeing of the fibers and are properly connected to the fibers. It has been determined that especially nanoscale TiO_2 and ZnO are efficient in absorbing and reflecting UV rays. This is a result of nanoparticles having larger surface areas per unit mass and unit. Because this situation increases the effectiveness of blocking UV rays. For small particles, the scattering of light is superior at about 0.1 wavelengths. According to Rayleigh scattering theory, scattering is proportional to λ-4. Accordingly, the best size of the particles should be 20-

40 nm to scatter UV rays between 200-400 nm. To block the UV light from the fabric produced from natural fibers, the nano technologically preferred method is the sol-gel method. For example, the TiO_2 layer is formed on the cotton fabric through the sol-gel method. This makes very good UV protection. This effect continues in 50 washes and ironings. Furthermore, 10-50 nm ZnO nanoparticles can be applied with this method on the cotton fabric for UV protection [55]. The protection function obtained with nanoparticles is permanent during usage. It does not show a significant decrease during the use, washing, or cleaning of textiles. In recent studies, the focus is on changing the surface properties of nanoparticles and reducing negative properties with the use of various combinations [57, 58].

SUPERHYDROPHOBIC TEXTILE APPLICATIONS OF NANOMATERIALS

The wettability of a solid surface is an important feature since different applications require different wettability of the surfaces. For the quantification of the wettability of a solid surface, it is possible to use contact angle measurement methods, which beyond simple measurement the contact angle between solid surface and droplet. Young's equation, presented in 1805, has been used as a basic equation. However, it has quite limited applications due to the necessity of a perfect surface, which is solid, purely flat, non-soluble, inert, and chemically homogenous. It is always possible for the real surfaces to have heterogenic chemical structure and surface roughness. The Wenzel Model, contrary to Young`s equation, considers the roughness of the surface but also assumes that the surface has a homogenous chemical structure. On the other hand, the Cassie-Baxter Model, unlike Wenzel, considers the flat surface but with chemical heterogeneity [59].

To define a surface as superhydrophobic, there are two basic requirements, (1) the surface must have water contact angle more than 150° and (2) water droplet must have low contact angle hysteresis [60], which defined as the difference between advancing and receding contact angles, typically lower than 10°, which indicates that droplet has good mobility on the surface. There are many examples of superhydrophobic surfaces in nature [61]. The most known one, *Lotus Leaf,* can easily repel water when it is raining with very small tilt angles due to an excellent combination of surface chemistry and roughness. There are many more examples from nature such as insects, plants, or even some living microorganisms.

Two main parameters contribute to superhydrophobicity, the first of them is the structure of the surfaces. Lotus leaf has a very tiny, hairy structure on it, which contributes the hydrophobicity. The second parameter, which affects the contact angle of the surface, is the chemical nature of the surface. This feature influences directly the surface tension of the surface, which has a critical role for wettability.

Although superhydrophobic textiles have important application areas such as self-cleaning oil-water separation, most of the conventional textiles do not exhibit superhydrophobicity. However, it is possible to impart superhydrophobicity to textile materials by modification of surface free energy, increasing surface roughness or surface etching methods. Specifically, designed nanomaterials treated fabrics are one of the most popular and easiest ways to produce superhydrophobic textiles. Non-fluorinated alkylsilanes or fluorosilanes functionalized SiO_2 nanoparticles treated textiles comprise the majority of the literature. These nanoparticles are used to impart superhydrophobicity to different types of textiles including cotton, polyester, polyamide, etc By the application of modified SiO_2 nanoparticles, both lower surface energy, and higher roughness values are achieved. Layer by layer assembly [62], spin coating [63], dip coating [64], and sol-gel techniques [65] are widely used to apply the nanomaterials providing superhydrophobicity. Generally, additional functions are aimed to be imparted to provide superhydrophobicity with other functional properties such as antimicrobial activity, UV-protection, etc.

Superhydrophobic textiles have a growing trend and an important topic for both industry and academia. Although there are several commercialized superhydrophobic textile products in the market, there are still lots of gaps in the literature about the application of nanomaterial for superhydrophobic textiles.

ELECTROMAGNETIC INTERFERENCE SHIELDING TEXTILE APPLICATIONS OF NANOMATERIALS

Most of the electronic devices, power lines, transformers, and wireless technology generates electromagnetic waves which may be harmful to human health with long exposure times. Although there are some commercialized EMI shielding materials on the market such as metal plates, they are generally suffering from some problems, stiffness heaviness, etc. Textiles are mostly considered as flexible, lightweight, low cost, and easy-to-produce materials.

These properties can make it possible to produce much better electromagnetic interference (EMI) shielding materials than that the commercialized ones.

The ratio between the incoming power ($P_{incoming}$) and the outgoing power ($P_{outgoing}$) gives the total efficiency (SE) of an EMI shielding material (Eq. 2) [66] and it is expressed in decibels (dB). It can be measured with a Vector Network Analyzer (VNA) using a coaxial measurement system.

$$SE = 10\log\left(\frac{P_{incoming}}{P_{outgoing}}\right) \quad \text{Eq. (2)}$$

After measuring the S11 and S21 values of an EMI shielding material using a vector network analyzer, the absorbance can be calculated from the following formula (Eq. 3),

$$P_{Absorbance} = 1 - (S21_{Transmittance})^2 - (S11_{Reflectance})^2 \quad \text{Eq. (3)}$$

When an electromagnetic wave strikes an EMI shielding material, there are three different mechanisms for attenuation:

1. Reflection
2. Absorption
3. Multiple reflections

During the absorption, electromagnetic energy converts into thermal energy. High dielectric materials or magnetic materials are used for EMI absorbers while the conductive materials and surfaces are used as EMI reflectors. A combination of absorption and reflection mechanisms provides the best efficiencies. However, the efficiency of an EMI shielding material not only depends on the materials used, but also the thickness, the amount and frequencies of the EM waves, and the homogeneity of EMI shielding materials.

Textile materials are known as electrically insulators and the conventional textiles are not suitable for EMI shielding applications. However, the EMI shielding effect can be achieved easily by using nanotechnology. Magnetic nanomaterials, metallic nanoparticles, and carbon-based nanomaterials are the most known materials to impart EMI shielding properties to the conventional textile structures.

Magnetic particles seem one of the best options to provide EMI shielding efficiency. In 2019, Yu et al., [67] produced EMI shielding Polyimide fabric using magnetic f-$NiFe_2O_4$ particles with a conductive polymer, polyaniline. The layer by layer method was used to prepare this three-phase heterostructure with high EMI absorption ability. They obtained around 42 dB of efficiency at the frequency of 12.5 GHz. In another study, Montazer et al., [68] prepared magnetic nickel-plated polyester fabric to achieve EMI shielding efficiency with an antibacterial activity using a cost-effective method, electroless plating. They obtained a significant efficiency at the frequencies between 8.4 and 12.4 GHz with some other functionalities such as electrical conductivity, magnetization, and higher tensile strain.

The first 2D material, graphene, is another promising candidate for EMI shielding applications. Its advantages such as high conductivity, flexibility easy-to-produce techniques, and compatibility with conventional textiles process make it possible to use graphene in functional textiles. In 2019, Fu et al., [69] used a continuous dyeing process to produced graphene oxide coated cotton fabrics and green reduction of GO was carried out by using ascorbic acid. It was found that the increasing amount of reduced graphene on cotton fabric provided better electrical conductivity as well as EMI shielding efficiency. In another study, Gupta et al., [70] prepared firstly the

Zinc oxide coated cotton fabric using the sol-gel method, and then they spray-dried the graphene to produce graphene/zinc oxide layered coating. It was reported that the prepared sample had an absorbance dominant EMI shielding mechanism due to the synergetic effect of highly dielectric ZnO and highly conductive graphene.

Metal nanocoatings are other favorable materials for EMI shielding applications. In 2020, Esen et al., [71] produced Al and Zn coated cotton/elastane fabrics using thermal vacuum evaporation technique and it was reported that Al coated samples have lower EMI shielding efficiencies than Zn coated samples. Aktas et al., [72] prepared metal nanoparticle decorated polyacrylonitrile textiles using electroless deposition technique and the EMI shielding efficiencies investigated in X and P bands. It was found that the EMI efficiency strongly dependent on deposition time and around 99.99% of attenuation achieved.

EMI shielding textiles have a high possibility of commercialization due to their wide application areas from protective clothes to military applications. Although there are some commercialized products in the market [73], it is believed that potential market share is not achieved yet.

Wrinkle Resistive Textile Applications of Nanomaterials

Wrinkling is the undesired bending deformation on the fabric and gives an aesthetically unpleasant appearance. In the crease resistance tests, the ability of the fabrics to return to their first form is measured when the pressure effect is removed after creasing under a certain pressure. The fiber elements that displace as a result of effects such as force, moisture content, force application time and temperature, create new side bonds. In this case, wrinkling occurs. Wrinkle resistance is particularly demanded in woven outerwear (e.g., suits, trousers, jackets, blouses, skirts, shirts, raincoats, and plain cellulosic fabrics). The fibers that are hard and have less bending ability wrinkle more than others. Also, wrinkle marks are more difficult to recover. Besides, wrinkling disappears quickly in flexible fibers. Fiber diameter and shape are also effective in wrinkle resistance. Fibers having round cross-section are more resistant to bending and folding. Yarn twist is also effective in the wrinkle resistance of fabrics. Low twist yarns are very easy to replace fiber in case of bending and cannot return to their original positions when the pressure is removed. Fibers in moderately twisted yarns have little or no mobility. Therefore, such yarns tend to return to their original positions. In very high twisted yarns, they tend to maintain the bending caused by tensions on the fiber or yarns, and therefore their wrinkle resistance is not as good as the moderately twisted yarns. Wrinkle resistance of thick fabrics produced from thick yarns is better. Fabrics with high fabric density tend to wrinkle more than sparse fabrics. Since the movement capability of the fabrics in the twill or satin weave construction is more, the wrinkles of plain fabrics produced by the same fabric density and yarn count are easier to recover. Knitted fabrics have better wrinkle resistance than woven fabrics since they have better flexing capabilities. Moreover, finishing processes applied to the fabric also affect the wrinkle resistance. The wrinkle resistance property of fabrics is primarily related to the ability of the fiber stretch. Wool and PES fibers have high stretching ability. For this reason, wrinkles formed in the fabrics produced from these fibers recover quickly and well. The wrinkle resistance properties of synthetics are generally better. However, fabrics produced from cotton, linen, regenerated cellulose (viscose), etc. fibers are sensitive to creasing [74].

Wrinkle resistance can be improved by finishing processes applied to these fabrics which are not good wrinkle resistance. The mobility of fiber elements is restricted by finishing. Even if they move, they are prevented from forming new side bonds in that position. Thus, the wrinkle resistance of the fibers under the crease force is increased. The first of the wrinkle-resistive materials used for this purpose are resin-forming substances. They react with each other and form a polymer structure (resin), filling the easy penetration (amorphous) regions of the fibers. The second is reactant substances that are bonded to the cellulose fibers over free-OH groups and bonded to them by a covalent bond. In both cases, the anti-crease effect is achieved by preventing the mobility of the fiber elements and the formation of new side bonds. Because the fiber elements displaced under the influence of the creasing force cannot approach each other or the reactive groups have already bonded.

However, wrinkle-resistant finishing processes have several drawbacks. Firstly, chlorine retention problems can be said. If the fabrics that have undergone wrinkle-resistance finishing treatment are maintained with cleaning agents containing chlorine (chlorinated water) and then ironed, chloramine causes the fabric yellowness and damages in the subsequent ironing process. The damaging substances to fiber are substances hypochlorous acid and active oxygen or hydrochloric acid released by its breakdown. Chlorine retention is no longer a major problem since the use of chlorine for cleaning has decreased. Today, still important problems in wrinkle-resistant finishing are the release of formaldehyde and the decrease in fabric strength. Also, problems with water intake, dyeing, and breathability are observed in resin applications. The most important drawbacks of classical wrinkle-resistant materials are that they release formaldehyde. Since reactive groups in classical wrinkle-resistant materials are formed by adding formaldehyde, formaldehyde is released both during work, storage, and on the fabric. Today, there are legal limits values of the released formaldehyde amounts for the working, working conditions, and each product group. The decrease in the breaking strengths as a result of the wrinkle-resistant treatment is due to the principle of the process, that is, the restriction of the mobility of the fiber elements. As the fiber elements, whose movement is restricted, do not combine and force together under the effect of a breakout force, it is inevitable to decrease the break strength as a result of such a process. The fact that the bonds are smooth and homogeneous has relatively some positive effects. The decrease in friction resistance is caused by the amount of chemical accumulated on the surfaces with migration during drying. If working is done by preventing migration with different working methods (less float application etc.), the decreases of friction resistance can be minimized. Due to the excessive decrease in the breaking and friction strength of the fabric as a result of the crease-free process, it is possible that such fabrics will suffer sewing damage during sewing in the garment if no precautions are taken (when there is no effective softener in the prescription) [75].

In recent years due to these restrictions, nanotechnology applications have become prominent to improve the properties of fabrics that have lower wrinkle resistance. In these applications, nano-TiO_2 and nano-SiO_2 are generally used. Nano-TiO_2 is used as an adjuvant for carboxylic acid and this is done under UV light. Due to its large surface area, nano $Ti0_2$ can fill the amorphous region of cellulose, and therefore the presence of nano-$Ti0_2$ in the fiber probably restricts the molecular motion of the cellulose. On the other hand, nano- SiO_2 is used as an adjuvant for maleic anhydride. In the silica nanoparticles, a covalent bond is formed between the -Si-OH and the hydroxyl group of cellulose, and the density of this bond plays a decisive role [55, 76-79]. Some studies focused on reducing negative features with the use of various nanoparticle combinations [80]. In recent years, studies that increase the wrinkle resistance with chitosan nanoparticle treatments come to the forefront. However, some studies have reported that nano-chitosan applications can cause yellowness [81, 82].

Conclusion and Future Trends

As seen, nanomaterials have great potential to prepare functional surfaces and materials. The technical textile industry needs to produce functional textile products with high added value to stay in the competition. Nanotechnology is a promising opportunity for the technical textiles and makes it possible to produce textiles with a high level of functionalities for the possible applications in protective clothes, military purposes, aviation industry, and home textiles. Metal oxides, carbon-based materials, and polymeric nanoparticles are used to prepare nanomaterial functionalized textiles. There are two main concerns about using nanomaterials. The first of them is the durability of the effect. Nowadays, scientists present different methods to enhance durability. However, more and more studies must be carried out because the cost of the functional textiles will be affordable only when it is possible to use them for a longer time. The second concern of using nanomaterials is their potential to expose the human body due to their extremely small sizes. Thus, it is an important research topic to keep the nanomaterials on the textiles during usage and prevent their migration to the human body as well as the environment. It is also strongly believed that bio-based nanomaterials are going to be much popular in the future and environmental effects of nanomaterials are going to be considered.

ACKNOWLEDGMENTS

This work was supported by The Scientific and Technological Research Council of Turkey (TUBITAK), The Department of Science Fellowships and Grant Programs (BIDEB).

REFERENCES

[1] Standardization, I. O. f., *Nanotechnologies–Vocabulary–Part 2: Nano-objects*. ISO/TS 80004-2: 2015, 2015.

[2] Pitkethly, M. J., Nanomaterials - the driving force. *Materials Today*, 2004. 7(12): p. 20-29.

[3] Chen, X., et al., Nanomaterials for renewable energy production and storage. *Chemical Society Reviews*, 2012. 41(23): p. 7909-7937.

[4] Kamyshny, A. and S. Magdassi, Conductive nanomaterials for printed electronics. *Small*, 2014. 10(17): p. 3515-3535.

[5] Gao, J. and B. Xu, Applications of nanomaterials inside cells. *Nano Today*, 2009. 4(1): p. 37-51.

[6] Peters, R. J., et al., Nanomaterials for products and application in agriculture, feed and food. *Trends in Food Science & Technology*, 2016. 54: p. 155-164.

[7] Harifi, T. and M. Montazer, A review on textile sonoprocessing: A special focus on sonosynthesis of nanomaterials on textile substrates. *Ultrasonics sonochemistry*, 2015. 23: p. 1-10.

[8] Alongi, J., F. Carosio, and G. Malucelli, Current emerging techniques to impart flame retardancy to fabrics: An overview. *Polymer Degradation and Stability*, 2014. 106: p. 138-149.

[9] Hajipour, M. J., et al., Antibacterial properties of nanoparticles. *Trends in biotechnology*, 2012. 30(10): p. 499-511.

[10] Montazer, M. and S. Seifollahzadeh, Enhanced self-cleaning, antibacterial and UV protection properties of nano TiO2 treated textile through enzymatic pretreatment. *Photochemistry and photobiology*, 2011. 87(4): p. 877-883.

[11] Gashti, M. P., A. Almasian, and M. P. Gashti, Preparation of electromagnetic reflective wool using nano-ZrO2/citric acid as inorganic/organic hybrid coating. *Sensors and Actuators A: Physical*, 2012. 187: p. 1-9.

[12] Cui, H. W., K. Suganuma, and H. Uchida, Highly stretchable, electrically conductive textiles fabricated from silver nanowires and cupro fabrics using a simple dipping-drying method. *Nano Research*, 2015. 8(5): p. 1604-1614.

[13] Ibrahim, N. A., *Nanomaterials for antibacterial textiles, in Nanotechnology in diagnosis, treatment and prophylaxis of infectious diseases*. 2015, Elsevier. p. 191-216.

[14] Sun, G., *Antimicrobial textiles*. 2016: Woodhead Publishing.

[15] Radetic, M., Functionalization of textile materials with silver nanoparticles. *Journal of Materials Science*, 2013. 48(1): p. 95-107.

[16] Rai, M., A. Yadav, and A. Gade, Silver nanoparticles as a new generation of antimicrobials. *Biotechnology advances*, 2009. 27(1): p. 76-83.

[17] Lee, H., S. Y. Yeo, and S. H. Jeong, Antibacterial effect of nanosized silver colloidal solution on textile fabrics. *Journal of Materials Science*, 2003. 38(10): p. 2199-2204.

[18] Pohle, D., et al., Antimicrobial properties of orthopaedic textiles after in-situ deposition of silver nanoparticles. *Polymers and Polymer Composites*, 2007. 15(5): p. 357-363.

[19] Vigneshwaran, N., et al., Functional finishing of cotton fabrics using silver nanoparticles. *Journal of Nanoscience and Nanotechnology*, 2007. 7(6): p. 1893-1897.

[20] Durán, N., et al., Antibacterial effect of silver nanoparticles produced by fungal process on textile fabrics and their effluent treatment. *Journal of biomedical nanotechnology*, 2007. 3(2): p. 203-208.

[21] Hadad, L., et al., Sonochemical deposition of silver nanoparticles on wool fibers. *Journal of applied polymer science*, 2007. 104(3): p. 1732-1737.

[22] Ki, H. Y., et al., A study on multifunctional wool textiles treated with nano-sized silver. *Journal of Materials Science*, 2007. 42(19): p. 8020-8024.

[23] Dastjerdi, R., M. Montazer, and S. Shahsavan, A new method to stabilize nanoparticles on textile surfaces. *Colloids and Surfaces A: Physicochemical and Engineering Aspects*, 2009. 345(1-3): p. 202-210.

[24] Jeong, S. H., S. Y. Yeo, and S. C. Yi, The effect of filler particle size on the antibacterial properties of compounded polymer/silver fibers. *Journal of Materials Science*, 2005. 40(20): p. 5407-5411.

[25] Dubas, S. T., P. Kumlangdudsana, and P. Potiyaraj, Layer-by-layer deposition of antimicrobial silver nanoparticles on textile fibers. *Colloids and Surfaces A: Physicochemical and Engineering Aspects*, 2006. 289(1-3): p. 105-109.

[26] Zhang, Y. Y., et al., Durable antimicrobial cotton textiles modified with inorganic nanoparticles. *Cellulose*, 2016. 23(5): p. 2791-2808.

[27] Çakır, B. A., et al., Synthesis of ZnO nanoparticles using PS-b-PAA reverse micelle cores for UV protective, self-cleaning and antibacterial textile applications. *Colloids and Surfaces A: Physicochemical and Engineering Aspects*, 2012. 414: p. 132-139.

[28] Lim, S. H. and S. M. Hudson, Review of chitosan and its derivatives as antimicrobial agents and their uses as textile chemicals. *Journal of Macromolecular Science-Polymer Reviews*, 2003. C43(2): p. 223-269.

[29] Ali, S. W., S. Rajendran, and M. Joshi, Synthesis and characterization of chitosan and silver loaded chitosan nanoparticles for bioactive polyester. *Carbohydrate Polymers*, 2011. 83(2): p. 438-446.

[30] Petkova, P., et al., Sonochemical coating of textiles with hybrid ZnO/chitosan antimicrobial nanoparticles. *ACS applied materials & interfaces*, 2014. 6(2): p. 1164-1172.

[31] Weil, E. A. and S. V. Levchik, Flame retardants in commercial use or development for polyolefins. *Journal of Fire Sciences*, 2008. 26(1): p. 5-43.

[32] Norouzi, M., Y. Zare, and P. Kiany, Nanoparticles as Effective Flame Retardants for Natural and Synthetic Textile Polymers: Application, Mechanism, and Optimization. *Polymer Reviews*, 2015. 55(3): p. 531-560.

[33] Wang, X., et al., Carbon-family materials for flame retardant polymeric materials. *Progress in Polymer Science*, 2017. 69: p. 22-46.

[34] Kausar, A., *7 - Flame retardant potential of clay nanoparticles, in Clay Nanoparticles*, G. Cavallaro, R. Fakhrullin, and P. Pasbakhsh, Editors. 2020, Elsevier. p. 169-184.

[35] White, L. A., Preparation and thermal analysis of cotton-clay nanocomposites. *Journal of Applied Polymer Science*, 2004. 92(4): p. 2125-2131.

[36] Vargas, A. F., et al., Influence of fiber-like nanofillers on the rheological, mechanical, thermal and fire properties of polypropylene An application to multifilament yarn. *Composites Part a-Applied Science and Manufacturing*, 2010. 41(12): p. 1797-1806.

[37] Porter, D., E. Metcalfe, and M. J. K. Thomas, Nanocomposite fire retardants - A review. *Fire and Materials*, 2000. 24(1): p. 45-52.

[38] Qin, H. L., et al., Flame retardant mechanism of polymer/clay nanocomposites based on polypropylene. *Polymer*, 2005. 46(19): p. 8386-8395.

[39] Song, P., et al., Fabrication of dendrimer-like fullerene (C-60)-decorated oligomeric intumescent flame retardant for reducing the thermal oxidation and flammability of polypropylene nanocomposites. *Journal of Materials Chemistry*, 2009. 19(9): p. 1305-1313.

[40] Beyer, G., Short communication: Carbon nanotubes as flame retardants for polymers. *Fire and Materials*, 2002. 26(6): p. 291-293.

[41] Sang, B., et al., Graphene-based flame retardants: a review. *Journal of Materials Science*, 2016. 51(18): p. 8271-8295.

[42] Zhang, W. J., G. Camino, and R. J. Yang, Polymer/polyhedral oligomeric silsesquioxane (POSS) nanocomposites: An overview of fire retardance. *Progress in Polymer Science*, 2017. 67: p. 77-125.

[43] Lam, Y. L., C. W. Kan, and C. W. M. Yuen, Effect of Titanium Dioxide on the Flame-Retardant Finishing of Cotton Fabric. *Journal of Applied Polymer Science*, 2011. 121(1): p. 267-278.

[44] Moafi, H. F., A. F. Shojaie, and M. A. Zanjanchi, Flame-retardancy and photocatalytic properties of cellulosic fabric coated by nano-sized titanium dioxide. *Journal of thermal analysis and calorimetry*, 2011. 104(2): p. 717-724.

[45] Lam, Y. L., C. W. Kan, and C. W. M. Yuen, Flame-Retardant Finishing in Cotton Fabrics Using Zinc Oxide Co-Catalyst. *Journal of Applied Polymer Science*, 2011. 121(1): p. 612-621.

[46] Abd El-Hady, M. M., A. Farouk, and S. Sharaf, Flame retardancy and UV protection of cotton based fabrics using nano ZnO and polycarboxylic acids. *Carbohydrate Polymers*, 2013. 92(1): p. 400-406.

[47] Liang, Y., et al., Synthesis and performances of Fe2O3/PA-6 nanocomposite fiber. *Materials Letters*, 2007. 61(14-15): p. 3269-3272.

[48] Dogan, M. and E. Bayramli, Effect of Boron Phosphate on the Mechanical, Thermal and Fire Retardant Properties of Polypropylene and Polyamide-6 Fibers. *Fibers and Polymers*, 2013. 14(10): p. 1595-1601.

[49] Laachachi, A., et al., Effect of Al2O3 and TiO2 nanoparticles and APP on thermal stability and flame retardance of PMMA. *Polymers for Advanced Technologies*, 2006. 17(4): p. 327-334.

[50] Paul, R., *Functional finishes for textiles: improving comfort, performance and protection*. 2014: Elsevier.

[51] Akaydin, M., İ. Yüksel, and N. S. Kurban, Measurement and evaluation of permeability of UV rays in cotton knitted fabrics. *Tekstil ve Konfeksiyon*, 2009. 19(3): p. 212-217.

[52] Khan, A., et al., A review of UV radiation protection on humans by textiles and clothing. *International Journal of Clothing Science and Technology*, 2020.

[53] Stanford, D. G., K. E. Georgouras, and M. T. Pailthorpe, Rating clothing for sun protection: current status in Australia. *Journal of the European Academy of Dermatology and Venereology*, 1997. 8(1): p. 12-17.

[54] Seventekin, N., *Textile Chemistry*. 2004: TEKAUM Publications. 110.

[55] Köksal, F. and R. Köseoğlu, *Nanoscience and Nanotechnologhy*. 2014: Nobel academic publishers.

[56] Schindler, W. D. and P. J. Hauser, *Chemical finishing of textiles*. 2004: Elsevier.

[57] Li, C., R. Li, and X. Ren, Inorganic-organic Hybrid Nanoparticles and Their Application on PET Fabrics for UV Protection. *Fibers and Polymers*, 2020. 21(2): p. 308-316.

[58] Bouazizi, N., et al., Development of new composite fibers with excellent UV radiation protection. *Physica E: Low-dimensional Systems and Nanostructures*, 2020. 118: p. 113905.

[59] Khodaei, M., Introductory Chapter: Superhydrophobic Surfaces-Introduction and Applications, in *Superhydrophobic Surfaces-Fabrications to Practical Applications*. 2019, IntechOpen.

[60] Ma, M. and R. M. Hill, Superhydrophobic surfaces. *Current opinion in colloid & interface science*, 2006. 11(4): p. 193-202.

[61] Samaha, M. A., H. V. Tafreshi, and M. Gad-el-Hak, Superhydrophobic surfaces: From the lotus leaf to the submarine. *Comptes Rendus Mecanique*, 2012. 340(1-2): p. 18-34.

[62] Zhao, Y., et al., Superhydrophobic cotton fabric fabricated by electrostatic assembly of silica nanoparticles and its remarkable buoyancy. *Applied Surface Science*, 2010. 256(22): p. 6736-6742.

[63] Shaban, M., F. Mohamed, and S. Abdallah, Production and characterization of superhydrophobic and antibacterial coated fabrics utilizing ZnO nanocatalyst. *Scientific reports*, 2018. 8(1): p. 1-15.
[64] Wu, L., et al., Mimic nature, beyond nature: facile synthesis of durable superhydrophobic textiles using organosilanes. *Journal of Materials Chemistry B*, 2013. 1(37): p. 4756-4763.
[65] Xue, C. H., et al., Superhydrophobic cotton fabrics prepared by sol–gel coating of TiO2 and surface hydrophobization. *Science and Technology of Advanced Materials*, 2008.
[66] Zhang, L., S. Bi, and M. Liu, *Lightweight Electromagnetic Interference Shielding Materials and Their Mechanisms, in Electromagnetic Materials and Devices*. 2018, IntechOpen.
[67] Wang, Y., W. Wang, and D. Yu, Three-phase heterostructures f-NiFe2O4/PANI/PI EMI shielding fabric with high Microwave Absorption Performance. *Applied Surface Science*, 2017. 425: p. 518-525.
[68] Moazzenchi, B. and M. Montazer, Click electroless plating of nickel nanoparticles on polyester fabric: Electrical conductivity, magnetic and EMI shielding properties. *Colloids and Surfaces a-Physicochemical and Engineering Aspects,* 2019. 571: p. 110-124.
[69] Islam, M. D. Z., et al., Continuous dyeing of graphene on cotton fabric: Binder-free approach for electromagnetic shielding. *Applied Surface Science,* 2019. 496.
[70] Gupta, S., et al., Reduced graphene oxide/zinc oxide coated wearable electrically conductive cotton textile for high microwave absorption. *Composites Science and Technology*, 2020. 188.
[71] Esen, M., et al., Investigation of electromagnetic and ultraviolet properties of nano-metal-coated textile surfaces. *Applied Nanoscience*, 2020. 10(2): p. 551-561.
[72] Akman, O., et al., Magnetic metal nanoparticles coated polyacrylonitrile textiles as microwave absorber. *Journal of Magnetism and Magnetic Materials*, 2013. 327: p. 151-158.
[73] Maity, S., et al., Textiles in electromagnetic radiation protection. *Journal of Safety Engineering*, 2013. 2(2): p. 11-19.

[74] Ozdil, N., *Physical Quality Control Methods in Fabrics*. 2003: TEKAUM Publications. 120.

[75] Coban, S., *General Textile Finishing Process*. Vol. 10. 1999: TEKAUM Publications.

[76] Yanilmaz, M., & Karakaş, H. ... 18(81). 35-41., Nanotechnology Applications for High Performance Textiles. *The Journal of Textiles and Engineer*, 2011. 18(81): p. 35-41.

[77] Ul-Islam, S. and B. S. Butola, *Nanomaterials in the Wet Processing of Textiles*. 2018: John Wiley & Sons.

[78] Yuen, C., et al., Using nano-tio 2 as co-catalyst for improving wrinkle-resistance of cotton fabric. *Surface review and letters*, 2007. 14(04): p. 571-575.

[79] Hezavehi, E., S. Shahidi, and P. Zolgharnein, Effect of dyeing on wrinkle properties of cotton cross-linked by Butane Tetracarboxylic Acid (BTCA) in presence of Titanium Dioxide (TiO2) Nanoparticles. *Autex Research Journal*, 2015. 15(2): p. 104-111.

[80] Noorian, S. A., N. Hemmatinejad, and A. Bashari, One-pot synthesis of Cu2O/ZnO nanoparticles at present of folic acid to improve UV-protective effect of cotton fabrics. *Photochemistry and Photobiology*, 2015. 91(3): p. 510-517.

[81] Lu, Y. H., et al., Preparation of chitosan nanoparticles and their application to Antheraea pernyi silk. *Journal of applied polymer science*, 2010. 117(6): p. 3362-3369.

[82] Tripathi, R., et al., Development of multifunctional linen fabric using chitosan film as a template for immobilization of in-situ generated CeO2 nanoparticles. *International journal of biological macromolecules*, 2019. 121: p. 1154-1159.

In: Challenges and Opportunities … ISBN: 978-1-53618-770-0
Editor: Wallace G. Tarrant

Chapter 3

TEXTILE INDUSTRY AND SHIFT WORK

Isabel S. Silva [*] ***and Daniela Costa***
School of Psychology, University of Minho, Braga, Portugal

ABSTRACT

The use of shift work is a practice adopted by many organizations, covering a significant number of workers. For example, according to Eurostat (2020), in 2019, 18.4% of workers in the 28 Member States of the European Union worked shifts; in Portugal, the value is slightly higher: 19.7%. On the other hand, there are significant differences in the use of this work schedule between sectors of activity. This is prevalent, for example, in the field of health, commerce and hospitality or in some industrial sectors (Eurofound, 2016).

In general, shift work, especially that involving night shifts, has been associated with several negative consequences for workers and organizations. This chapter is dedicated to reviewing the main consequences associated with shift work for shift workers (health and family and social life) and for organizations (safety and performance). This review will focus on studies carried out in an industrial context in general, and in the textile sector in particular. The chapter ends with recommendations on the management of shift work schedules, where the

* Corresponding Author's Email: isilva@psi.uminho.pt.

importance of flexible practices in this management will be emphasized. In this topic, it will also be presented the results of empirical studies carried out under the scientific coordination of the first author, in the Portuguese industrial context, including the textile sector, where the relationship between negative effects associated with shift work (e.g., health) and the adoption of certain management practices by organizations is analyzed (e.g., worker participation in the management of working hours).

Keywords: shift work, textile industry, health, work/life balance, safety, performance, flexible work arrangements

INTRODUCTION

The economic (e.g., globalization of the economy), technological (e.g., expansion of information technology) and demographic (e.g., woman's entry into the labor market) changes that have taken place in the recent decades have significantly changed the functioning of the labor market and families (Presser 2009, 1778). For example, according to Olivetti and Petrongolo (2016), between 1950 and 2010, female employment (considering the 15 to 64 years age group) increased by about 40% in the United States of America (USA) and 50% in Portugal. With the aim of responding to some of these changes, the organizations often increase operational time; working up to 24 hours a day, and 7 days a week. This increase in working time leads to the emergence of non-conventional working hours; working hours different from the conventional time, characterized by working on working days from 8:00 am to 5:00 pm. Costa (2003, 83) argued that the non-conventional schedules encompass hourly modalities, such as shift work, part-time, and flexible hours among others.

The textile industry, according to the *Global Textile Industry – growth, trends and forecast (2020-2025)* report by Mordor Intelligence (2020), is constantly growing with China, European Union (EU), USA and India as the main players of this market. Within these, China appears to be the world's leading producer and exporter of textiles and raw clothing, while Portugal is one of the top five countries in the EU's textile industry, such as Germany,

Spain, France, and Italy (Mordor Intelligence 2020). Taking into account another publication, but this time national in scope, in 2016, Portugal represented 4.43% of the turnover of the textile industry and 5.38% of the clothing industry in the EU (Cenit 2017, 43). In terms of characterization of the textile and clothing sectors in Portugal, in 2016, there were more than 12,200 companies (28.76% of the textile sector and 71.24% of the clothing sector) that represented 18% of manufacturing companies and 1% of companies of the country (Cenit 2017, 41). According to the same source, the textile sector employed over 13% of workers in the manufacturing industry, while the clothing sector employed about 7% of the workers. Finally, in the geographical terms, the North of Portugal is the region that concentrates these two industrial subsectors, and in terms of merchandise exports, these subsectors were responsible for 9.5% of the exports (Cenit 2017, 46).

The next point in this chapter will address the main consequences associated with shift work at three levels: health, family and social life, and organizational level. Within the scope of this review, studies carried out in an industrial context were privileged. The second part of the chapter will be dedicated to the presentation of recommendations on the management of shift work schedules based on the literature consulted as well as the research carried out under the scientific coordination of the first author in the Portuguese industrial context on shift work.

Shift Work

Shift work is defined by Costa (1997, 89) as "a way of organization of daily working hours, wherein different teams work in succession to cover more or all of the 24 h." In 2019, according to the European Union Statistics Office – Eurostat, 18.4% of workers in the 28 Member States of the EU worked in shifts, while Portugal had a percentage of shift work of 19.7 (Eurostat 2020). According to the same office, the member states with the highest percentage of shift work were Slovenia (36%), Croatia (33.9%), and

Poland (31.4%). Conversely, the member states with the lowest percentage of shift work were France (6.6%), Belgium (7.1%), and Denmark (8.4%).

Industrial Sector

Fiz Perez et al. (2019, 159) found out that in the European continent, until half of the last century, shift work was a strategy used in sectors such as health, transport, communications, or public security. However, with changes in the society, other sectors (e.g., industry, road construction, commerce) started using these working hours to respond to the demands of the market. In 2015, in the EU, the sectors where shift work was most common were health (41%), transport (33%), industry (28%), and commerce and hospitality (28%) (Eurofound 2016, 59). Specifying in the Portuguese context, according to the National Statistics Institute - INE (2020), in 2019, the *industry, construction, energy and water sector* comprised of 1,212,400 workers (24.68%), of whom 143,700 worked in shifts (11.85%); 122,800 practiced night work (10.13%); 328,000 worked on Saturday (27.05%), and 104,700 worked on Sunday (8.64%). Also, according to this institute, men are the vast majority (78.91%) of shift workers in this secondary sector.

Consequences of Shift Work

Like any other schedule, shift work can have advantages and disadvantages for workers. The advantages that are mainly associated with shift work are the temporal reorganization (i.e., availability to perform certain types of activities that would not be possible during conventional hours) and economic advantages, given that as a rule, a shift allowance is associated with a shift work (Silva 2008, 363; West et al. 2012). On the other hand, the disadvantages that have been associated with shift work are mainly focused on three areas: health, family and social life, and organizational level (Costa 1997, 90; Silva 2008). In the next subsections, a synthesis of the main impacts that have been studied in each of these areas will be made.

In the study by Carneiro and Silva (2015, 155), carried out by the research group of the first author, some advantages and disadvantages of shift work were identified. "Availability to deal with matters" and "flexibility of schedules" were the most reported advantages, while "interference with family and friends" and "sleep interference" were the most reported disadvantages by workers.

Health

Over the years, the literature (e.g., Åkerstedt 2003; Costa 1997, 91) has reinforced the negative impacts that shift work has on workers, especially on their health, such as sleep, digestive and/or cardiovascular diseases among others.

Sleep problems have been the most studied impacts of shift work that concern the workers more (Åkerstedt 2003). Problems such as sleep quality, disorders, and excessive sleepiness have been among the most reported problems in the literature (e.g., Narciso et al. 2014, 203; Øyane et al. 2013, 4). For example, Lim et al. (2018), using a sample of Malaysian industry workers, found out that shift work was associated with poorer sleep quality, greater latency to fall asleep, shorter sleep duration, sleep disturbances, and greater difficulties in keeping awake during day.

The literature has found evidence of the existence of an association between the practice of different working hours from the conventional working hours, namely shift work and night work, and *psychological problems,* such as depression and anxiety (e.g., Ardekani et al. 2008; Lee et al. 2017). In the study by Kim et al. (2016), developed in the South Korean automotive industry, it was observed that shift workers had a higher depression rate than day workers. Also, in the study of Ajmera, Satia, and Singh (2016, 1368), in the Indian industrial context, the authors found that shift workers (rotation between evening, night, and day shift) and night workers (00:00-08:00) had higher levels of anxiety and stress than workers in the day shift (08:00-17:00) and evening shift (16:00-00:00).

In addition to sleep and psychological problems, shift work also appears to cause difficulties for systems fundamental to the functioning of the human body, such as the *cardiovascular system* (e.g., Haupt et al. 2008; Wang et al. 2018). In the study by Asare-Anane et al. (2015, 3), with a sample of 113 shift workers vs. 87 day workers in a Ghanaian industry, the authors found a positive association between shift work and risk factors for the development of cardiovascular diseases, such as BMI, and fasting blood glucose among others.

The most reported *gastrointestinal problems* over the years have been peptic ulcers, gastritis, and abdominal pain among others (e.g., Knutsson and Bøggild 2010; Swanson et al. 2016, G197). In the context of the South Korean electronics industry, Lee et al. (2020) observed that shift workers had higher rates of gastritis and reported more gastrointestinal symptoms, including indigestion, than day workers.

At the *metabolic system* level, shift work seems to be negatively associated with diseases, such as diabetes and obesity (e.g., Di Lorenzo et al. 2003, 1355; Li et al. 2019; Shaker et al. 2018, 93). For example, in the study by Rabanipour et al. (2019, 196), conducted with workers from Iranian steel companies (1,683 shift workers vs. 1,380 day workers), the results indicated that shift workers had a higher risk of being overweight. Also, Ika et al. (2013, 28), with a sample of 475 male workers from a Japanese manufacturing company, found out that shift workers were at an increased risk of developing diabetes. The study of these impacts of shift work on the metabolic system has gained greater relevance due to the severity of consequences, which a dysfunction in this system can cause, for individuals, such as strokes or myocardial infarctions. Also, the *endocrine system*, responsible for hormone production, seems to be affected by shift work (e.g., Magrini et al. 2006; Shaker et al. 2018, 93). For example, in the meta-analysis carried out by Coppeta et al. (2020), the authors found that Thyroid Stimulating Hormone (TSH) values were significantly higher in night workers than in day workers, resulting in some concern if we think that high

TSH values may mean thyroid gland disorders that can consequently cause more serious illnesses like heart problems.

In the recent years, *oncological problems* (e.g., breast cancer, colorectal cancer) have also been associated with shift work schedules, mainly involving the night shift (e.g., Rao et al. 2015, 2822; Yuan et al. 2018, 36). The literature has suggested that the potential risk of developing carcinomas in night shift workers comes in part from circadian interruption and low levels of melatonin produced by these workers (Haus and Smolensky 2013; Wei et al. 2020). However, several meta-analyzes carried out in the recent years still check for inconsistencies in this association (Kamdar et al. 2013; Rivera-Izquierdo et al. 2020, 7). For example, in the study by Li et al. (2015, 6), no evidence was found that shift work increased the risk of breast cancer in a sample of workers in the Chinese textile industry.

Finally, associations between shift work and the *female reproductive system* have also been studied, mainly in the process of pregnancy and the menstrual cycle (Costa 1997, 92; Cai et al. 2019; Lin et al. 2011). For example, Lin et al. (2011, 132), using a sample of workers from a Chinese semiconductor factory, found evidence that shift work was associated with a decrease in the number of pregnancies and a reduction in fetal growth. In the meta-analysis by Stocker et al. (2014, 102), working on a non-conventional schedule was associated with a higher rate of infertility and menstrual interruption, and working at night was associated with an increase in miscarriage in early pregnancy.

Family and Social Life

At the family and social level, the impacts of shift work tend to arise due to the lack of synchrony between the hours practiced by shift workers and those of other members of the household and society. In this sense, as some authors defend (e.g., Camerino et al. 2010, 1116; Costa 2003, 85; Handy 2010, 33; Mauno, Ruokolainen, and Kinnunen 2015, 89), shift work tends to be associated with a greater work-family conflict (WFC), which can translate into marital and/or parental conflicts. For example, Davis et al.

(2008, 1001), synthesizing the results of two studies conducted in the North American context, verified the association between night work and greater conjugal instability and work-family spillover as compared to daytime and weekend hours. The results of the study also indicated that working at the weekend caused more stressors for the work-family relationship than working during the week. The study by Tuttle and Garr (2012, 265), using a sample of more than 2,000 North American workers, also revealed that shift work was associated with an increase in the WFC as compared to day work. In the same sense, when workers had some kind of control over their working hours, WFC decreased. In a more recent study, in the Australian context, Zhao et al. (2020, 14) also observed that working at irregular hours increased the risk of WFC.

In terms of *marital relationships*, the literature (e.g., Maume and Sebastian 2012, 485; Minnotte, Minnotte, and Bonstrom 2015, 27) has confirmed the existence of these problems in association with shift work. For example, Wight, Raley, and Bianchi (2008, 264) found out that married parents who worked different hours than the conventional working hours spent less time with their spouse, slept less, and watched television for less time as compared to parents who worked during conventional hours. Gracia and Kalmijn (2016, 409), in the Spanish context, found a negative association between night work and family time and time spent as a couple.

Parental relationships can also suffer from the practice of non-conventional working hours. As some authors (e.g., Begall, Mills, and Ganzeboom 2015, 971; Han and Fox 2011; Li et al., 2014) have noticed, over the last few years, the shift work practiced by one of the parents can affect not only the parents themselves, but also their children. Still in relation to the study by Wight, Raley, and Bianchi (2008, 260) mentioned earlier, the authors found out that parents who worked in the evening hours spent less time with children as compared to day workers, and parents who worked the night shift spent more time overall and more alone time with the children. In a slightly different sense, Zilanawala et al. (2017, 718), in the British context, found out that children where both parents worked at non-conventional hours had a higher BMI growth trajectory than children where both parents worked at conventional hours.

At the *social level*, several authors (e.g., Baker, Ferguson, and Dawson 2003, 36; Martin, Wittmer, and Lelchook 2011, 914) have demonstrated the existence of a negative association between shift work and this dimension. For example, Craig and Brown (2014) found out that weekend work was associated with a decrease in workers' leisure time, and the majority of such workers was unable to recover this shared leisure time during the following week.

Organizations

Notwithstanding the impacts that shift work can have on workers, whether in terms of health or in terms of family and social life, organizations can also be affected, in terms of workers' safety as well as productivity, by the practice of these working hours, especially at night (Costa 1997, 93; Folkard and Tucker 2003, 99). For example, Costa (1997, 94) stated that the organizations started giving greater importance to shift work, when they realized that several historical disasters have occurred during the night shift, as was the case with the Exxon Valdez in the USA in 1989. In fact, Folkard and Tucker (2003, 97) found out that the risk of incidents at the workplace increased linearly clockwise (morning shift, afternoon shift, and night shift), with the night shift being the time with the highest risk of occurrence of an incident. In addition to the night shift problem, a 12-hour shift was more detrimental to the safety of workers than an eight-hour shift (Folkard and Tucker 2003, 98).

In the systematic review by Wagstaff and Sigstad (2011), the authors found in all 14 studies that shift work as well as long hours of work negatively influenced workers' *safety*. Also, Bergh, Shahriari, and Kines (2013, 406), using a sample of workers from two Swedish chemical companies, found that shift workers as compared to day workers scored significantly lower on the seven scales of the Nordic Occupational Safety Climate Questionnaire (NOSACQ-50): management of "safety priority and ability," "safety empowerment," "safety justice," "safety commitment," "safety priority and risk non-acceptance," "peer safety communication, learning, and trust in safety ability," and "trust in the efficacy of safety systems."

Like worker safety, *productivity* and *performance* have been negatively associated with shift work (e.g., Caruso 2014, 21; Folkard and Tucker, 2003, 96). For example, in the literature review by Dall'Ora et al. (2016, 14), work in rotating shifts was associated with worst job performance. Also, Ganesan et al. (2017), with a sample of Australian healthcare workers, found that night work was associated with more detrimental results in workers' attention and performance. Houboubi et al. (2017, 69), using a sample of 125 workers from an Iranian petrochemical company, found that shift workers had significantly lower average productivity scores than day workers.

In addition to the variables mentioned above, there have been studies that have assessed the relationship between shift work and other organizational variables, such as absenteeism and job satisfaction (e.g., Bernstrøm and Houkes 2020; Jaradat et al. 2017, 73). In terms of *absenteeism*, Morikawa et al. (2001, 396), in a sample of more than 2,500 workers from a Japanese factory, found that in the previous year, workers who had worked more than two third of working days during non-day hours had a higher incidence of sick leave than other workers. In turn, Jacobsen and Fjeldbraaten (2018), using workers from a Norwegian hospital, did not find direct effects of shift work on the absence of illness, but they found indirect effects through the mediation role of WFC and perceived health. On the other hand, in terms of *job satisfaction*, Ferri et al. (2016, 207) found that workers on rotating shifts had lower average scores as compared to day workers.

In short, shift work has been associated with several negative impacts on the lives of individuals, their families, and the organizations themselves, making it increasingly important to explore measures to prevent these difficulties. The second part of the chapter will be dedicated to the presentation of recommendations and intervention strategies that help minimize the impacts of shift work. Such recommendations will be made based on the literature and research carried out in the Portuguese context on adaptation to shift work.

Recommendations/Intervention Strategies

As seen initially, labor market demands tend to force organizations to work 24 hours a day, changing working hours, and consequently increasing non-conventional hours, such as shift work. However, the emergence of these schedules can have negative consequences for workers, their families, and organizations. As some authors have argued (e.g., Åkerstedt and Wrigth 2009; Demerouti et al. 2004), the most logical measure would be to extinguish shift work schedules, especially the night shift, but given the impossibility of doing so, strategies should be devised to help minimize their impacts. In general, we can say that such strategies are focused on an individual level (e.g., strategies aimed at sleep) or on an organizational context (e.g., design of shift systems).

Åkerstedt and Wrigth (2009, 9) synthesized some countermeasures to minimize the impacts of shift work on sleep, sleepiness, and fatigue. They, specifically, recommend clockwise rotation of shifts (i.e., morning, afternoon, and night) in order to decrease the number of successive night shifts, increase flexibility and thc influcncc of workcrs on thc managcmcnt of their own working hours, or educate workers about good sleep habits. In the same sense, Berger and Hoobs (2006) also presented some countermeasures to improve sleep and circadian rhythms: establish regular patterns of sleep, work, and leisure, avoid contact with blue lights after the end of a night shift, reduce all noise inside the room, avoid food, drinks, or other substances with caffeine before bedtime, do physical exercise, and schedule moments with the family that coincide with the schedules of both among others. More generally, Dhande and Sharma (2011, 267) made some recommendations, such as night work should be avoided, existence of a cafeteria 24 hours a day to provide hot and nutritious meals to workers of the night shift, workers should receive training to minimize the impacts of working hours, and rapid shift changes should be avoided among others. Smith and Eastman (2012) presented feasible countermeasures to address the impacts of the night shift, such as stimulants (e.g., caffeine) or naps before or during the night shift to increase attention and performance, and melatonin to promote sleep during the day. In turn, Hoboubi et al. (2017,

71), aiming at reducing occupational stress and increasing worker satisfaction and productivity, presented some measures, such as identification and optimization of factors that influence these variables, and adjustment to shift work.

The organizational context can also play an important role in the process of adaptation to work through the management practices of working hours that it adopts. For example, Moradi et al. (2014, 65) identified that nurses who voluntarily chose to work in shifts had significantly higher job satisfaction than those who were forced to work in shifts. Other studies (e.g., Krausz, Sagie, and Bidermann 2000; Fenwick and Tausig 2001; Brooks and Swailes 2002) have supported the relevance of the variable 'individual control over working time'. For example, the study by Fenwick and Tausig (2001, 1189) analyzed the effects of different types of working time arrangements (e.g., fixed day shift, fixed non-daily shift from Monday to Friday, rotating shifts) and the perception of working time control on various indicators on health and family life based on data from a US national survey. Overall, they found that workers with higher levels of working time control reported significantly fewer problems in both domains. Also, Camerino et al. (2010) found that the development of a preventive culture could be effective in reducing WFC and increasing well-being and job performance in workers assigned to shift work and night work.

Given the significance of the organizational context, this dimension has been studied in the research group under scientific guidance from the first author. Specifically, efforts have been made to comprehend the role of shift work management/adaptation practices, with the research carried out (e.g., Silva 2008; Silva and Prata 2015) pointing to different forms of this management by organizations. For example, while some organizations seek to involve workers in choosing the shifts they are going to work for, others make this decision unilaterally. Thus, in this context, a scale was developed by Silva (2008) to capture the perception of organizational support (POS) in the management of aspects related to working time. The original version of this scale consists of four items: "*The company does everything possible to place workers in the shifts they prefer,*" "*As a rule, when workers need to change shifts, the company responds favorably,*" "*The company takes into account the situation of workers when it is necessary to have personnel changes between shifts,*" and "*The company takes into account the workers' preferences for a particular shift when they are hired.*" Cronbach's alpha value was 0.84. This scale has been integrated into studies on the effects of shift work, and the results of two studies are presented below (Silva 2008; Ferreira and Silva 2013); both studies were conducted in Portuguese textile companies.

The first study developed by Silva (2008) was conducted involving 247 shift workers, while the second study by Ferreira and Silva (2013) included a sample of 122 shift workers. The average age in both studies was found to be 36.11 (SD = 11.61) and 38.93 (SD = 9.40) respectively. The samples in both studies consisted mainly of men, married or living in a de facto union with children. In terms of seniority in working hours, the two studies had an average of more than 10 years, with the first study having an average of 11.2 years (SD = 12.79) and the second study with an average of 10.94 years (SD = 8.31).

Both studies aimed at assessing the relationship between the effects of working hours and POS. Specifically, Silva (2008) studied the relationship between POS and sleep problems, digestive problems, life outside the company, and satisfaction with working hours. In turn, Ferreira and Silva (2013) analyzed the interaction between POS and life outside the company, family life, conjugal life, and satisfaction with working hours. Table 1 shows the results of the correlations between the effects of shift work evaluated in each study and the POS.

Table 1. Correlations between the effects of shift work and POS

Effects of shift work	**Silva (2008) (N = 247)**	**Ferreira and Silva (2013) (N = 122)**
Sleep problems	-.19**	---
Digestive problems	-.31***	---
Life outside the company	.36***	.41***
Family life	---	.32*
Conjugal life	---	.30**
Satisfaction with working hours	.28***	.30**

* p < .05; ** p < .01; *** p < .001.

As can be seen in Table 1, the results obtained indicate a statistically significant association between the effects evaluated in each study and the POS as well as the theoretically expected direction. Thus, considering the first study, sleep problems and digestive problems were negatively associated with POS, i.e., the higher the POS, the lower the sleep problems and digestive problems. On the other hand, satisfaction with life outside the company and satisfaction with working hours were positively associated with the POS, i.e., the higher the POS, the better the perception of the articulation between working hours and life outside the company and the greater the satisfaction with their own working hours. In the second study, the POS was positively associated with all the effects assessed, i.e., the higher the POS, the greater the satisfaction with life outside the company, family life, and conjugal life in addition to working hours.

In general, the results obtained support the role that the organizational context may have in adaptation to shift work; in particular, the way it designs and implements management practices for specific aspects relating to working hours (e.g., selection and placement of workers on shifts), or the way people are managed when shifts are reorganized. Taken as a whole, such practices seek to enhance the adjustment between individual preferences/needs of shift workers and organizational needs through the involvement of workers in this management. The possibility of this involvement, on the part of workers, translates for us at least two advantages. On the one hand, it makes it easier to accommodate differences between workers from a situational point of view (e.g., reconciling with their children's school schedules; spouse's working hours). On the other hand, it also allows more easily accommodating differences in the individual/biological domain (e.g., daytime type, that is the preferences of individuals regarding the hours of activities and rest, morning/evening).

CONCLUSION

In this chapter, we had the opportunity to present, in the first part, the negative impacts of shift work at the level of the worker's life spheres (health and family and social life) as well as the organizational level (e.g., safety). The second part of the chapter was dedicated to the presentation and discussion of possibilities for intervention in order to minimize such impacts. In this discussion, the role that the organizational context can play in the management of aspects of working time was privileged, including empirical evidence based on two companies in the textile sector. Despite such an approach, this context can still play an important role in managing the potential difficulties imposed by such work regimes, especially at night. This management can be performed through the resources it makes available (e.g., food service or work-home transport) besides respecting the ergonomic recommendations in the design of shift systems (e.g., avoid abrupt changes between shifts, or not have enough time to recover between shift changes).

REFERENCES

Ajmera, P., Satia, H. K., and Singh, M. 2016. "Impact of shift work schedules on levels of stress, anxiety and work life balance in BPO employees." *International Journal of Recent Advances in Multidisciplinary Research*, 3:1367-70.

Åkerstedt, T. 2003. "Shift work and disturbed sleep/wakefulness." *Occupational Medicine* 53:89-94. doi:10.1093/occmed/kqg046.

Åkerstedt, T., and Wright, K. P. 2009. "Sleep loss and fatigue in shift work and shift work disorder." *Sleep Medicine Clinics* 4:257-71. doi:10.1016/j.jsmc.2009.03.001.

Ardekani, Z. Z., Kakooei, H., Ayattollahi, S. M., Choobineh, A., and Seraji, G. N. 2008. "Prevalence of mental disorders among shift work hospital nurses in Shiraz, Iran." *Pakistan Journal of Biological Sciences: PJBS* 11:1605-9. doi:10.3923/pjbs.2008.1605.1609.

Asare-Anane, H., Abdul-Latif, A., Ofori, E. K., Abdul-Rahman, M., and Amanquah, S. D. 2015. "Shift work and the risk of cardiovascular disease among workers in cocoa processing company, Tema." *BMC research notes* 8:798. doi:10.1186/s13104-015-1750-3.

Baker, A., Ferguson, S., and Dawson, D. 2003. "The perceived value of time: Controls versus shiftworkers." *Time & Society* 12:27-39. doi:10.1177/0961463X03012001444.

Begall, K., Mills, M., and Ganzeboom, H. B. G. 2015. "Non-standard work schedules and childbearing in the Netherlands: A mixed-method couple analysis." *Social Forces* 93:957-88. doi:10.1093/sf/sou110.

Berger, A. M., and Hobbs, B. B. 2006. "Impact of shift work on the health and safety of nurses and patients." *Clinical Journal of Oncology Nursing* 10:465-71. doi:10.1188/06.CJON.465-471.

Bergh, M., Shahriari, M., and Kines, P. 2013. "Occupational safety climate and shift work." *Chemical Engineering Transactions* 31:403-8. doi:10.3303/CET1331068.

Bernstrøm, V. H., and Houkes, I. 2020. "Shift work and sickness absence at a Norwegian hospital: A longitudinal multilevel study." *Occupational and Environmental Medicine* 77. doi:10.1136/oemed-2019-106240.

Brooks, I., and Swailes, S. 2002. "Analysis of the relationship between nurse influences over flexible working and commitment to nursing." *Journal of Advanced Nursing* 38:117-26. doi:10.1046/j.1365-2648.2002.02155.x.

Cai, C., Vandermeer, B., Khurana, R., Nerenberg, K., Featherstone, R., Sebastianski, M., and Davenport, M. H. 2019. "The impact of occupational shift work and working hours during pregnancy on health outcomes: A systematic review and meta-analysis." *American Journal of Obstetrics and Gynecology* 221:563-76. doi:10.1016/j.ajog.2019.06.051.

Camerino, D., Sandri, M., Sartori, S., Conway, P.M., Campanini, P., and Costa, G. 2010. "Shiftwork, work-family conflict among Italian nurses, and prevention efficacy." *Chronobiology International* 27:1105–23.

Carneiro, L., and Silva, I. S. 2015. "Trabalho por turnos e suporte do contexto organizacional: Um estudo num centro hospitalar." *International Journal on Working Conditions* 9:142-60. ["Shift work and organizational support: A study in a hospital." *International Journal on Working Conditions* 9:142-60].

Caruso, C. C. 2014. "Negative impacts of shiftwork and long work hours." *Rehabilitation Nursing* 39:16-25. doi:10.1002/rnj.107.

Cenit. 2017. *Têxtil e Vestuário no Contexto Nacional e Internacional: Publicação Anual 2017*. Centro de Inteligência Têxtil. [*Textiles and Clothing in the National and International Context: Annual Publication 2017*. Textile Intelligence Center].

Coppeta, L., Di Giampaolo, L., Rizza, S., Balbi, O., Baldi, S., Pietroiusti, A., and Magrini, A. 2020. "Relationship between the night shift work and thyroid disorders: A systematic review and meta-analysis." *Endocrine Regulations* 54:64-70. doi:10.2478/enr-2020-0008.

Costa, G. 1997. "The problem: Shiftwork." *Chronobiology International* 14:89-98. doi:10.3109/07420529709001147.

Costa, G. 2003. "Shift work and occupational medicine: An overview." *Occupational Medicine* 53:83-8. doi:10.1093/occmed/kqg045.

Craig, L., and Powell, A. 2011. "Non-standard work schedules, work-family balance and the gendered division of childcare." *Work, Employment & Society* 25:274-91. doi:10.1177/0950017011398894.

Dall'Ora, C., Ball, J., Recio-Saucedo, A., and Griffiths, P. 2016. "Characteristics of shift work and their impact on employee performance and wellbeing: A literature review." *International Journal of Nursing Studies* 57:12-27. doi:10.1016/j.ijnurstu.2016.01.007.

Davis, K. D., Benjamin Goodman, W., Pirretti, A. E., and Almeida, D. M. 2008. "Nonstandard work schedules, perceived family well-being, and daily stressors." *Journal of Marriage and Family* 70:991-1003. doi:10.1111/j.1741-3737.2008.00541.x.

Demerouti, E., Geurts, S. A., Bakker, A. B., and Euwema, M. 2004. "The impact of shiftwork on workhome conflict, job attitudes and health." *Ergonomics* 47:987-1002. doi:10.1080/00140130410001670408.

Dhande, K. K., and Sharma, S. 2011. "Influence of shift work in process industry on workers' occupational health, productivity, and family and social life: An ergonomic approach." *Human Factors and Ergonomics in Manufactoring & Service Industries* 21:260-8. doi:10.1002/hfm.20231.

Di Lorenzo, L., De Pergola, G., Zocchetti, C., L'Abbate, N., Basso, A., Pannacciulli, N., Cignarelli, M., Giorgino, R., and Soleo, L. 2003. "Effect of shift work on body mass index: Results of a study performed in 319 glucose-tolerant men working in a Southern Italian industry." *International Journal of Obesity* 27:1353-8.

Eurofound. 2016. *Sixth European Working Conditions Survey*. Luxembourg: Publications Office of the European Union.

Eurostat. 2020. "Employees working shifts as a percentage of the total of employees, by sex and age (%)." Accessed July 13. https://ec.europa.eu/eurostat/web/products-datasets/-/lfsa_ewpshi.

Fenwick, R., and Tausig, M. 2001. "Scheduling stress – Family and health outcomes of shift work and schedule control." *American Behavioral Scientist* 44:1179-98. doi:10.1177/00027640121956719.

Ferreira, A. I., and Silva, I. S. 2013. "Trabalho em turnos e dimensões sociais: Um estudo na indústria têxtil." *Estudos de Psicologia* 18:477-85. ["Shiftwork and social dimension: A study in the textile industry." *Studies of Psychology* 18:477-85] doi:10.1590/S1413-294X2013000300008.

Ferri, P., Guadi, M., Marcheselli, L., Balduzzi, S., Magnani, D., and Di Lorenzo, R. 2016. "The impact of shift work on the psychological and physical health of nurses in a general hospital: A comparison between rotating night shifts and day shifts." *Risk Management and Healthcare Policy* 9:203-11. doi:10.2147/RMHP.S115326.

Fiz Perez, J., Traversini, V., Fioriti, M., Taddei, G., Montalti, M., and Tommasi, E. 2019. "Shift and night work management in European companies." *Calitatea* 20:157-65.

Folkard, S., and Tucker, P. 2003. "Shift work, safety and productivity." *Occupational Medicine* 53:95-101. doi:10.1093/occmed/kqg047.

Ganesan, S., Magee, M., Stone, J. E., Mulhall, M. D., Collins, A., Howard, M. E., Lockley, S. W., Rajaratnam, S. M. W., and Sletten, T. L. 2019. "The impact of shift work on sleep, alertness and performance in healthcare workers." *Scientific Reports* 9:1-13. doi:10.1038/s41598-019-40914-x.

Gracia, P., and Kalmijn, M. 2016. "Parents' family time and work schedules: The split-shift schedule in Spain." *Journal of Marriage and Family* 78:401-15. doi:10.1111/jomf.12270.

Han, W. J., and Fox, L. E. 2011. "Parental work schedules and children's cognitive trajectories." *Journal of Marriage and Family* 73:962-80. doi:10.1111/j.1741-3737.2011.00862.x.

Handy, J. 2010. "Maintaining family life under shiftwork schedules: A case study of a New Zealand petrochemical plant." *New Zealand Journal of Psychology* 39:29-37.

Haupt, C. M., Alte, D., Dörr, M., Robinson, D. M., Felix, S. B., John, U., and Völzke, H. 2008. "The relation of exposure to shift work with atherosclerosis and myocardial infarction in a general population." *Atherosclerosis* 201:205-11. doi:10.1016/j.atherosclerosis.2007.12.059.

Haus, E. L., and Smolensky, M. H. 2013. "Shift work and cancer risk: potential mechanistic roles of circadian disruption, light at night, and sleep deprivation." *Sleep Medicine Reviews* 17:273-84. doi:10.1016/j.smrv.2012.08.003.

Hoboubi, N., Choobineh, A., Ghanavati, F. K., Keshavarzi, S., and Hosseini, A. A. 2017. "The impact of job stress and job satisfaction on workforce productivity in an Iranian petrochemical industry." *Safety and Health at Work* 8:67-71. doi:10.1016/j.shaw.2016.07.002.

Ika, K., Suzuki, E., Mitsuhashi, T., and Takao, S. 2013. "Shift work and diabetes mellitus among male workers in Japan: does the intensity of shift work matter?" *Acta Medica Okayama* 67:25-33.

INE. 2020. "População empregada (Série 2011 - N.º) por Sexo, Sector de atividade económica (CAE Rev. 3) e Tipo de horário de trabalho; Anual." ["Employed population (Series 2011 - No.) by Sex, Economic sector (CAE Rev. 3) and Type of working hours; Annual"]. Accessed May 26. https://www.ine.pt/xportal/xmain?xpid=INE&xpgid=ine_indicadores&indOcorrCod=0006153&contexto=bd&selTab=tab2.

Jacobsen, D. I., and Fjeldbraaten, E. M. 2018. "Shift work and sickness absence—the mediating roles of work–home conflict and perceived health." *Human Resource Management* 57:1145-57. doi:10.1002/hrm.21894.

Jaradat, Y. M., Nielsen, M. B., Kristensen, P., and Bast-Pettersen, R. 2017. "Shift work, mental distress and job satisfaction among Palestinian nurses." *Occupational Medicine* 67:71-4. doi:10.1093/occmed/kqw128.

Kamdar, B. B., Tergas, A. I., Mateen, F. J., Bhayani, N. H., and Oh, J. 2013. "Night-shift work and risk of breast cancer: a systematic review and meta-analysis." *Breast Cancer Research and Treatment* 138:291-301.

Kim, B. I., Chung, T. H., Jeon, Y. J., Jang, J. H., Jin, H. M., and Cho, Y. J. 2016. "Relationship between Shift Work and Depression in Male Workers in a Car Production Plant." *Korean Journal of Family Practice* 6:356-61. doi:10.21215/kjfp.2016.6.4.356.

Knutsson, A., and Bøggild, H. 2010. "Gastrointestinal disorders among shift workers." *Scandinavian Journal of Work, Environment & Health* 36:85-95.

Krausz, M., Sagie, A., and Bidermann, Y. 2000. "Actual and preferred work schedules and scheduling control as determinants of job-related attitudes." *Journal of Vocational Behavior* 56:1-11. doi:10.1006/jvbe.1999.1688.

Lee, S., Chae, C. H., Park, C., Lee, H. J., and Son, J. 2020. "Relationship of shift work with endoscopic gastritis among workers of an electronics company." *Scandinavian Journal of Work, Environment & Health* 46:161-7. doi:10.5271/sjweh.3862.

Lee, A., Myung, S. K., Cho, J. J., Jung, Y. J., Yoon, J. L., and Kim, M. Y. 2017. "Night shift work and risk of depression: meta-analysis of observational studies." *Journal of Korean medical science* 32:1091-6. doi:10.3346/jkms.2017.32.7.1091.

Li, J., Johnson, S. E., Han, W., Andrews, S., Kendall, G., Stradzins, L., and Dockery, A. 2014. "Parents' nonstandard work schedules and child well-being: A critical review of the literature." *Journal of Primary Prevention* 35:53-73. doi:10.1007/s10935-013-0318-z.

Li, W., Chen, Z., Ruan, W., Yi, G., Wang, D., and Lu, Z. 2019. "A meta-analysis of cohort studies including dose-response relationship between shift work and the risk of diabetes mellitus." *European Journal of Epidemiology* 34:1013–24.

Li, W., Ray, R. M., Thomas, D. B., Davis, S., Yost, M., Breslow, N., Gao, D. L., Fitzgibbons, E. D., Camp, J. E., Wong, E., Wernli, K. J., and Checkoway, H. 2015. "Shift work and breast cancer among women textile workers in Shanghai, China." *Cancer Causes & Control* 26:143-50. doi:10.1007/s10552-014-0493-0.

Lim, Y. C., Hoe, V. C., Darus, A., and Bhoo-Pathy, N. 2018. "Association between night-shift work, sleep quality and metabolic syndrome." *Occupational and Environmental Medicine* 75:716-23. doi:10.1136/oemed-2018-105104.

Lin, Y. C., Chen, M. H., Hsieh, C. J., and Chen, P. C. 2011. "Effect of rotating shift work on childbearing and birth weight: A study of women working in a semiconductor manufacturing factory." *World Journal of Pediatrics* 7:129-35. doi:10.1007/s12519-011-0265-9.

Magrini, A., Pietroiusti, A., Coppeta, L., Babbucci, A., Barnaba, E., Papadia, C., Iannaccone, U., Boscolo, P., Bergamaschi, E., and Bergamaschi, A. 2006. "Shift work and autoimmune thyroid disorders." *International Journal of Immunopathology and Pharmacology* 19:31-6.

Martin, J. E., Wittmer, J. L. S., and Lelchook, A. M. 2011. "Attitudes towards days worked where Sundays are scheduled." *Human Relations* 64:901-26. doi:10.1177/0018726710396248.

Maume, D. J., and Sebastian, R. A. 2012. "Gender, nonstandard work schedules, and marital quality." *Journal of Family and Economic Issues* 33:477-90. doi:10.1007/s10834-012-9308-1.

Mauno, S., Ruokolainen, M., and Kinnunen, U. 2015. "Work–family conflict and enrichment from the perspective of psychosocial resources: Comparing Finnish healthcare workers by working schedules." *Applied Ergonomics* 48:86-94. doi:10.1016/j.apergo.2014.11.009.

Minnotte, K. L., Minnotte, M. C., and Bonstrom, J. 2015. "Work– family conflicts and marital satisfaction among US workers: Does stress amplification matter?" *Journal of Family and Economic Issues* 36:21–33. doi:10.1007/s10834-014-9420-5.

Moradi, S., Farahnaki, Z., Akbarzadeh, A., Gharagozlou, F., Pournajaf, A., Abbasi, A. M., Omidi, L., Hami, M., and Karchani, M. 2014. "Relationship between shift work and Job satisfaction among nurses: A Cross-sectional study." *International Journal of Hospital Research* 3:63-8.

Mordor Intelligence. 2020. "Global Textile Industry – growth, trends and forecast (2020-2025)." Accessed June 15. https://www.mordor intelligence.com/industry-reports/global-textile-industry---growth-trends-and-forecast-2019---2024.

Morikawa, Y., Miura, K., Ishizaki, M., Nakagawa, H., Kido, T., Naruse, Y., and Nogawa, K. 2001. "Sickness absence and shift work among Japanese factory workers." *Journal of Human Ergology* 30:393-8. doi:10.11183/jhe1972.30.393.

Narciso, F. V., Teixeira, C. W., Silva, L. O., Koyama, R. G., Carvalho, A. N. S., Esteves, A. M., Tufik, S., and Mello, M. T. 2014. "Maquinistas ferroviários: trabalho em turnos e repercussões na saúde." *Revista Brasileira de Saúde Ocupacional* 39:198-209. ["Train drivers: Shiftwork and health impacts." *Brazilian Journal of Occupational Health* 39:198-209.] doi:10.1590/0303-7657000084113.

Olivetti, C., and Petrongolo, B. 2016. "The evolution of gender gaps in industrialized countries." *Annual Review of Economics* 8:405-34. doi:10.1146/annurev-economics-080614-115329.

Øyane, N. M., Pallesen, S., Moen, B. E., Åkerstedt, T., and Bjorvatn, B. 2013. "Associations between night work and anxiety, depression, insomnia, sleepiness and fatigue in a sample of Norwegian nurses." *PloS one* 8:1-7. doi:10.1371/journal.pone.0070228.

Presser, H. B. 1999. "Toward a 24-hour economy." *Science* 284:1778-9. doi:10.1126/science.284.5421.1778.

Rabanipour, N., Roohafza, H., Feizi, A., Tirani, S. A., and Sarrafzadegan, N. 2019. "Association between shift work and obesity in a large sample of Iranian steel industry workers." *Archives of Industrial Hygiene and Toxicology* 70:194-200. doi:10.2478/aiht-2019-70-3266.

Rao, D., Yu, H., Bai, Y., Zheng, X., and Xie, L. 2015. "Does night-shift work increase the risk of prostate cancer? A systematic review and meta-analysis." *OncoTargets and Therapy* 8:2817-26. doi:10.2147/OTT.S89769.

Rivera-Izquierdo, M., Martínez-Ruiz, V., Castillo-Ruiz, E. M., Manzaneda-Navío, M., Pérez-Gómez, B., and Jiménez-Moleón, J. J. 2020. "Shift Work and Prostate Cancer: An Updated Systematic Review and Meta-Analysis." *International Journal of Environmental Research and Public Health* 17:1-17. doi:10.3390/ijerph17041345.

Shaker, D., Samir, A., Zyada, F., El-Sharkawy, M., and Ekladious, S. M. 2018. "Impact of shift work on sleep problems, hormonal changes, and features of metabolic syndrome in a sample of Egyptian industrial workers: A cross-sectional study." *Middle East Current Psychiatry* 25:91-7. doi:10.1097/01.XME.0000526694.65550.93.

Silva, I. S. 2008. "Adaptação ao trabalho por turnos." PhD diss. em Psicologia do Trabalho e das Organizações, Universidade do Minho. ["Adaptation to shift work." PhD diss. in Work and Organizational Psychology, University of Minho].

Silva, I. S., and Prata, J. 2015. "Work schedules and human resource management: The case of shift work." In *Human Resource Management Challenges and Changes*, edited by C. Machado, and J. P. Davim, 67-93. New York: Nova Publishers.

Smith, M. R., and Eastman, C. I. 2012. "Shift work: Health, performance and safety problems, traditional countermeasures, and innovative management strategies to reduce circadian misalignment." *Nature and Science of Sleep* 4:111-32. doi:10.2147/NSS.S10372.

Stocker, L. J., Macklon, N. S., Cheong, Y. C., and Bewley, S. J. 2014. "Influence of shift work on early reproductive outcomes: A systematic review and meta-analysis." *Obstetrics & Gynecology* 124:99-110. doi:10.1097/AOG.0000000000000321.

Swanson, G. R., Gorenz, A., Shaikh, M., Desai, V., Kaminsky, T., Van Den Berg, J., Murphy, T., Raeisi, S., Fogg, L., Vitaterna, M. H., Forsyth, C., Turek, F., Burgess, H. J., and Keshavarzian, A. 2016. "Night workers with circadian misalignment are susceptible to alcohol-induced intestinal hyperpermeability with social drinking." *American Journal of*

Physiology-Gastrointestinal and Liver Physiology 311:G192-G201. doi:10.1152/ajpgi.00087.2016.

Tuttle, R., and Garr, M. 2012. "Shift work and work to family fit: Does schedule control matter?" *Journal of Family and Economic Issues* 33:261-71. doi:10.1007/s10834-012-9283-6.

Wagstaff, A. S., and Lie, J. A. S. 2011. "Shift and night work and long working hours-a systematic review of safety implications." *Scandinavian Journal of Work, Environment & Health* 37:173-85. doi:10.5271/sjweh.3146.

Wang, D., Ruan, W., Chen, Z., Peng, Y., and Li, W. 2018. "Shift work and risk of cardiovascular disease morbidity and mortality: A dose–response meta-analysis of cohort studies." *European Journal of Preventive Cardiology* 25:1293-302.

Wei, T., Li, C., Heng, Y., Gao, X., Zhang, G., Wang, H., Zhao, X., Meng, Z., Zhang, Y., and Hou, H. 2020. "Association between Night-Shift-Work and Level of Melatonin: Systematic Review and Meta-Analysis." Available at http://dx.doi.org/10.2139/ssrn.3537100.

West, S., Mapedzahama, V., Ahern, M., and Rudge, T. 2012. "Rethinking shiftwork: Mid-life nurses making it work." *Nursing Inquiry* 19:177-87. doi:10.1111/j.1440-1800.2011.00552.x.

Wight, V. R., Raley, S. B., and Bianchi, S. M. 2008. "Time for children, one's spouse and oneself among parents who work nonstandard hours." *Social Forces* 87:243–71. doi:10.1353/sof.0.0092.

Yuan, X., Zhu, C., Wang, M., Mo, F., Du, W., and Ma, X. 2018. "Night shift work increases the risks of multiple primary cancers in women: a systematic review and meta-analysis of 61 articles." *Cancer Epidemiology and Prevention Biomarkers* 27:25-40. doi:10.1158/1055-9965.EPI-17-0221.

Zhao, Y., Cooklin, A. R., Richardson, A., Strazdins, L., Butterworth, P., and Leach, L. S. 2020. "Parents' Shift Work in Connection with Work–Family Conflict and Mental Health: Examining the Pathways for Mothers and Fathers." *Journal of Family Issues* 1-29. doi:10.1177/0192513X20929059.

Zilanawala, A., Abell, J., Bell, S., Webb, E., and Lacey, R. 2017. "Parental nonstandard work schedules during infancy and children's BMI trajectories." *Demographic Research* 37:709-26. doi:10.4054/DemRes. 2017.37.22.

In: Challenges and Opportunities … ISBN: 978-1-53618-770-0
Editor: Wallace G. Tarrant

Chapter 4

LIFE-CYCLE ASSESSMENT AND INDUSTRY 4.0: AN INTEGRATED MODEL FOR THE TEXTILE AND APPAREL INDUSTRY

Adriana Yumi Sato Duarte[1,*]***,***
Regina Aparecida Sanches[2]***,***
Rayana Santiago de Queiróz[3]***,***
Fernando Soares de Lima[3] ***and Franco Giuseppe Dedini***[4]

[1]Faculty of Communication, Arts and Design,
Nossa Senhora do Patrocínio University, Salto, São Paulo, Brazil
School of Arts. Science and Humanities, University of São Paulo,
São Paulo, Brazil
Institute for Technological Research, São Paulo, Brazil
[2]School of Arts. Science and Humanities, University of São Paulo,
São Paulo, Brazil
[3]Institute for Technological Research, São Paulo, Brazil
[4]Facult of Mechanical Engineering, State University of Campinas,
Campinas, São Paulo, Brazil

[*] Corresponding Author's Email: adriana.duarte@ceunsp.edu.br.

Abstract

Each Industrial Revolution played a key role in the production-consumption pattern of textile products. Given the complexity of the Textile and Apparel Industry, the production systems have varying degrees of conceptual and technological innovation. Thus, the objective of this research is to present a tool that integrates Life-cycle Assessment and Industry 4.0. For this, the analysis was delimited to four production systems for the manufacturing of 100% combed cotton t-shirts: obtaining of textile fibers, spinning, knitting and garment confection. The LCA indicated a participation of all systems analyzed in the generation of negative impacts on human health, climatic, atmospheric and soil conditions, but the stage of transformation of fibers into wires presented the worst environmental indexes. From the technological monitoring, it was possible to infer that the conceptual novelty and complexity of configuration of the components are in an evolutionary process, in which there is automation of machine parts and processes. To achieve the Fourth Industrial Revolution, technological tools must be incorporated into both the production and consumption of textiles.

Keywords: textile and apparel industry, industry 4.0, life-cycle assessment, forecasting

Introduction

Textile production is intrinsically related to the origin of humanity. Clothing, as well as food and shelter, is considered a basic human need. In addition, it is a form of self-expression that becomes even more important for individuals and social groups [1].

The textile industry has played and still plays a major role in the Industrial Revolution. The textile and apparel industry include the obtaining of raw materials, spinning, weaving and knitting, dyeing and printing, production and sale [2].

Since the First Industrial Revolution, there have been modifications in the production-consumption and in the organization of society. In the contemporary competitive context, technological tools, the need for a rapid response in the decision-making process, the demand for functional products and the appropriate social and environmental responsibilities of organizations are fundamental principles for the development of new products [3].

The First Industrial Revolution allowed the mechanization of the productive process and stimulated consumption and the need for constant renewal of the products. The Second Industrial Revolution introduced new sources of energy, mass production and the revolution of consumer goods. The Third Industrial Revolution was characterized by the transition from analog to digital and by the segmentation of consumer markets. The Fourth Industrial Revolution, which is the one currently taking place, is based on the integrative and collaborative environment, on the adoption of Cyber-Physical Systems and on the active role of consumers in the production system [4, 5, 6].

Thus, this work aims to present a tool that associates the Life-Cycle Assessment and the Industry 4.0, analyzing the main innovations in the production system and machinery through the assessment by analyzing patents and technological forecasting and comparing the environmental impacts from different production systems. For this, the analysis was delimited to four production systems for the manufacturing of 100% combed cotton t-shirts: obtaining of textile fibers, spinning, knitting and garment confection.

RESEARCH SCENARIO

Industry 4.0

In the contemporary's competitive environment, the product development process includes technological tools, the need for rapid decision-making, analytical tools, high productivity and lower cost, demand for functional products, and social and environmental responsibilities. The consumer needs and desires are mostly related to four concepts: speedy, smart, slim and sustainability.

The latest 20 years, internet has changed the way people communicate. The first era (1995), internet was an integrated hypermedia, in the second era (2000) internet had a programming media approach, that changed to people's web service in the third era (2005), and the fourth era - which encompasses nowadays – represents a new level of organization and management of the entire value chain on the products' life cycle. Industry 4.0 is an integrative cyber-physical system based on modern control systems, embedded software systems and Internet addresses. This industrial revolution is based on improvement of brainwork, especially in engineering activities, and fast decision-making [7].

According to the German Academy of Science and Engineering (Deutsch Akademie der Technnikwissenschaft - Acatech), the Industry 4.0 is the next industrial revolution. The Industry 4.0 is not only a technical challenge but will also change the organizational structure of industries. For the first time, an industrial revolution is assessed a priori and not ex-post, which in other words means a prediction of what is to happen and not an assessment of what has passed [8].

This concept became public in 2011, during the Hanover Fair, where representatives of economics, politics and academia promoted the idea of strengthening the competitiveness of German industry. The federal government supported the initiative and announced the first recommendations for implementing the Industry 4.0 which were later published in 2013 [8].

Once the future working place will be digital and flexible, a company must be data-driven and open-minded to innovation. Regarding the labor skills, [9] argue that unlike the occurred in the previous Industrial Revolution – in which machines replaced workers – the Fourth Industrial Revolution calls for a greater integration between man and cyber-physical systems in a way that their individual skills can be fully utilized and they will act in a wider area of expertise. [10] describes this labor model as "self-programmable labor."

The Fourth Industrial Revolution forecasts a smart product and smart machines. A smart product contains information about its production processes, communicates with the production chain, and decides what steps to take. A smart machine predicts failure or quality problems, organizes the decision-making process and self-optimization [11].

In this context, all industrial elements will be equipped with data processing able to communicate with each other that will compose a more complex process management. The Fourth Industrial Revolution foresees the need for connection between all in-formation flows.

The main goal of Industry 4.0 is to improve the value chains among the product´s lifecycle. In this context, the improvement of industrial competitiveness is achieved by organizing and controlling value, new business models and networks creation process [6].

One interesting aspect is the change in the product lifecycle with the inclusion of sustainability as a design requirement. The increasing environmental awareness modifies the consumer behavior. Consumers are active actors in this scenario, and the purchase decision will be based on the initiatives of companies.

[12] suggest that the techno-logical tools will support the remote tracking of an entire production chain and the environmental impacts throughout the design process. [13] suggests the alignment of the Fourth Industrial Revolution with sustainability by adopting the 17 goals described by the United Nations (Sustainable Development Goals - SDGs) to be met by the year 2030.

Once ethics, environmental and social impacts are directly related to the production processes, and the K-waves emphasize this trend, they represent an ongoing movement that should be considered as an active part in the contemporary society [14]. The future of production-consumption paradigm as the end of advantages of low-cost labor in the production system, intensive use of ubiquitous technology and new production systems (mini-factories, digital factories, 3D printing).

The result of the 4th Industrial revolution will be the Smart Factory, where Cyber Physical Systems (CPS) and Internet of Things (IoT) are the key technologies for achieving the production goals. The Industrie 4.0 is focused on creating products, procedures and processes so that smart factories are able to manage complexity, interruptions and efficiency. In a smart factory, humans, machines and resources communicate with each other as easily as in a social network [12, 15].

Life-Cycle Assessment (LCA)

Products interact with the environment through energy and material flows at all phases of the production process, from the extraction of raw materials, manufacturing, transportation and distribution, use and maintenance, reuse and recycling, and finally, waste management and disposal [16]. In the contemporary society, the demand for sustainable products and the improvement in implementation of environmental laws by regulatory authorities clearly demonstrate a growing recognition of the importance of moving towards a more sustainable model of industrial production [17].

In 1983, the UN (United Nations) created the World Commission on Environment and Development, and a report published in 1987 originated the concept of sustainable development. In this report, entitled Our Common Future, sustainable development "meets the needs of the present without compromising the ability of future generations to meet their own needs" [18]. Despite the fact that this term has more than 200 definitions in academic literature, all concepts relate economy, society and ecological environment as a long-term strategy [19].

Several nations set targets and standards for mitigating the deleterious effects of climate change by reducing emissions of greenhouse gas effect, replacing the supply of traditional energy with renewable energy resources and increasing energy efficiency [20]. There are many tools to dimension qualitative and quantitative environment impacts to support the transition toward a new socioeconomic model. One example is the Life Cycle Assessment (LCA) which is defined according to ISO 14040 (2006) as a technique to identify and select opportunities to improve environmental indicators [21].

The product life cycle simulation has emerged as a promising field to reduce the gap between the estimated costs of the supply chain and design decisions [22]. [23] states that LCA is applied to investigate the environmental management along the production chain and to avoid problem shifting, i. e., to solve a problem without transferring it to the next stakeholder. [24] suggest a broader application to LCA by going beyond the company gate and including different actors to create a collaboration network.

LCA method is divided into four phases: goal and scope definition, inventory analysis, impact assessment and interpretation. The first phase includes the system boundary, the reasons for applying LCA, the intended audience and the functional unit. Life cycle inventory analysis quantifies inputs/outputs of a product system, including resources and air/water/land emissions. The third phase evaluates the environmental impacts by comparing the inventory analysis results and environmental impacts data available in scientific reports. Last phase, interpretation, involves conclusion and recommendations [21].

Textile and Apparel Industry

The Textile and Apparel Industry is growing in volume and productivity, however it is facing several problems with environmental and social impacts. Consumer education, durability, price, business models, technology development, infrastructure, government policy and international agreements are seen as key factors to ensure a sustainable future [25].

According to the WTO (World Trade Organization, 2015), in 2013 the global amount of exportations of textiles and garments was US$ 766 billion, China has been the leading exporter both in textiles (39%) and clothing (35%) while the European Union (EU) is the largest importer (38%). In the year 2020, this volume will be USS 851 billion [26].

The future trends on textile and apparel production is the increasing in the labor skill level, new production technologies, pressure from consumers towards an environmentally production, and international campaigns for better work conditions [25].

The growing demand for eco products has developed a new requirement: the eco labeling. The ecolabels are designed to inform consumers about the environmental and social conditions in which they were produced, so they can make the decision at the time of purchase; however, it seems that the concern for the environment may not be enough to motivate consumers to buy environmentally friendly products [27].

[28] proposes an improvement in textile and apparel industry by "improving quality and durability of the product itself, producing products that can be recycled easily or cheaply, producing in one location rather another, maintaining long-term relationships with supplier companies, paying above the minimum possible rates of pay, ensuring better working conditions for the employees of supplier companies, minimizing the use of energy/water/scarce resources, minimizing the pollution resulting from production, minimizing the environmental degradation resulting from the production of raw materials."

In the textile machinery, [29] suggested a new configuration that changes the conventional textile industry to a flexible factory in which machine communicates through IoT platform its re-al-time status and upcoming problems. In addition, the same IoT platform could guide the consumer through a mobile app for the use-phase (washing, e.g.,) and disposal, two critical phases in a textile life cycle assessment.

METHODS

The methodological procedures adopted in this research have a predominantly exploratory character, based on a case study which had as main aim the technological forecasting of the textile industry.

The delimitation of the research is the manufacturing of 100% combed cotton t-shirts. To this end, four production systems were determined: the obtaining of cotton fiber, spinning, knitting and garment confection.

The functional deployment described by [30] was applied for each production system. This technique relates inputs and outputs, which correspond to material/energy/information, functions and subfunctions.

Each system was detailed in subsystems, and each subsystem was associated with one or more components (machinery and/or machine elements).

The LCA was simulated in the openLCA program, based on data collected in literature. This software was developed in 2006 by the company GreenDelta and has been updated periodically. Thus, new results were generated, allowing the comparison of environmental impacts between the four production systems.

The technology forecasting model contemplated two methods: Assessment, with the aim of monitoring the development of the factors that can be agents of change, and Forecasting, which is the projection of technological changes.

The Technological Assessment was based on documents obtained in patent databases, including Brazilian and foreign patents, from 2011 to the present day. For this research, two databases were selected: an international one in English (Google Patents) and a Brazilian one in Portuguese (*Instituto Nacional da Propriedade Industrial* – INPI). Finally, the Forecasting was performed using historical information and the theoretical modeling of trends.

Results and Discussion

This section was divided into three main topics: specification of production systems and components, LCA and technological assessment and forecasting.

Production Systems and Components

The textile production systems adopted in this research are the processing of cotton fiber, spinning, knitting and garment confection that correspond to the following systems: Obtain Textile Fibers, Turn Fibers into Yarn, Turn Yarn into Fabrics and Turn Fabrics into Ready-to-Wear Clothing, as shown in Figure 1.

However, due to the complexity of the systems, a new ramification was created to specify the subsystems.

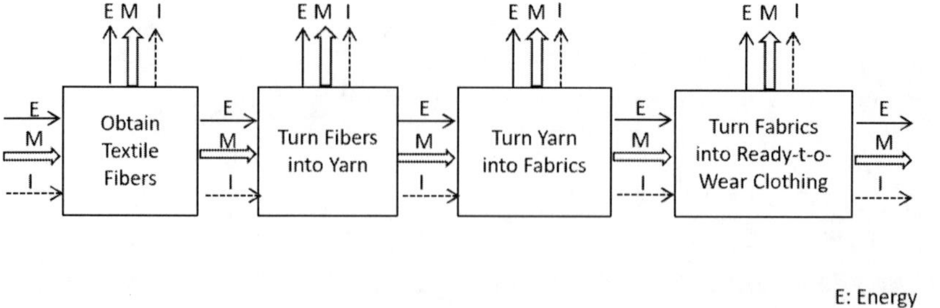

Figure 1. Four systems (Authors).

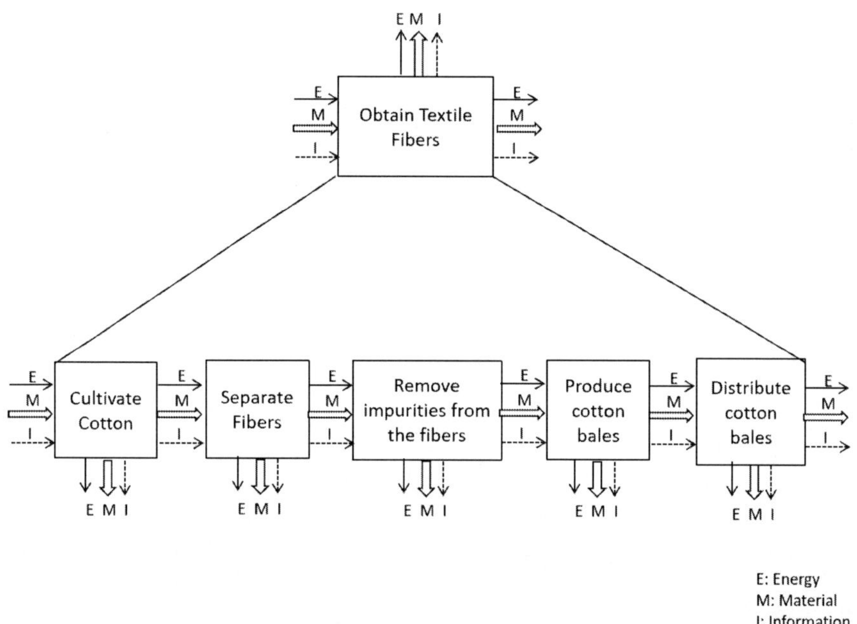

Figure 2. Obtain textile fibers (Authors).

The "Obtain Textile Fibers" system encompasses the steps of cotton cultivation, post-harvest separation of fibers, removal of impurities from the fibers, production of cotton bales and distribution of the bales for marketing, as illustrated in Figure 2.

The cultivation of cotton requires a wide range of agricultural machinery that perform tasks of soil preparation, distribution of seeds in the soil at a specific depth and spacing, and harvesting of the products.

The separation of fibers consists in the use of the Stripper Harvester and the Cotton Gin, machines which harvest and separate the cottonseed fibers, respectively. The elimination of impurities from the fibers is made with a Cotton Cleaning Machine.

The cotton baling is performed by the elements Cotton Condenser, Module Feeder and Bailing Press. The Condenser starts the process of removing of air from the fibers, forming a blanket, which will be taken by the Feeder to the Press so that the bale can finally be formed. For the distribution of cotton bales, the transportation conduit and the assessment system will be needed.

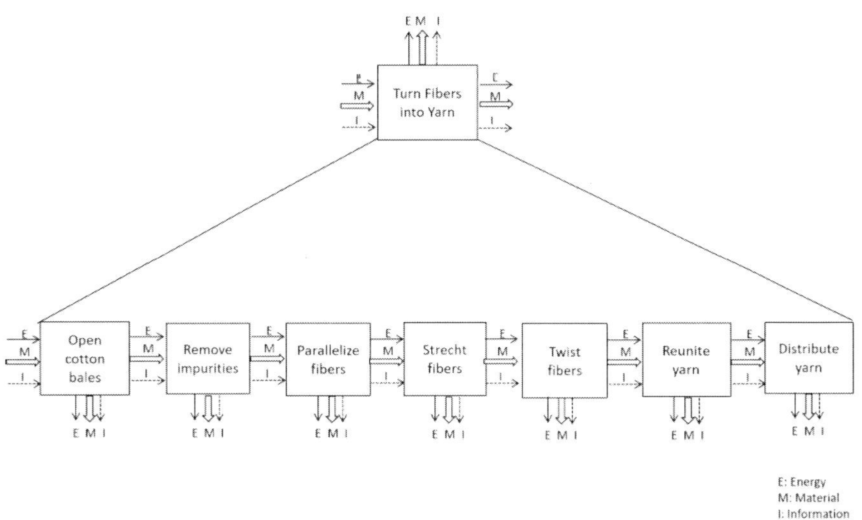

Figure 3. Turn fibers into yarn (Authors).

The "Turn Fibers into Yarn" system encompasses the opening of the bales, the elimination of impurities such as bark and leaves, parallelization, stretching and twisting of the fibers, gathering of the yarn produced and its subsequent distribution, as shown in Figure 3.

The "Turn Fibers into Yarn" system encompasses the opening of the bales, the elimination of impurities such as bark and leaves, parallelization, stretching and twisting of the fibers, gathering of the yarn produced and its subsequent distribution, as shown in Figure 3.

The opening of the cotton bales is performed in the Blow Room, where they are opened for separating the fibers in order to facilitate the formation of the yarn. The carding machine performs the individual rearrangement of the fibers to eliminate impurities and parallelize them.

In the Drawing machine, the fibers are first stretched, and the weight/unit length is determined. The Pre-Combing Drawing Frame, the Lap former machine and the Comber are the set of machines responsible for parallelization and for the removing of short fibers. The Rover and Spinner stretch the fibers and spin them together. The distribution of the finished yarn is conducted through a transportation conduit and assessment system.

The production system "Turn Yarn into Fabric" has the knitted fabric as basis. The knitted fabric can be formed by a single yarn (weft knitting) or several yarns (warp knitting), loops being the basic constructive element of this fabric.

The specification of the components of this function is related to the constructive elements of the single-cylinder circular knitting machine, which produces single Jersey fabrics intended to produce t-shirts. In this machine, the formation of the fabric corresponds to the action of the components: yarn guide (responsible for guiding the yarn onto the needles) and steel needle bed. The movement of the needles for the formation of the fabric is performed by the set of components: cam, needles, and sinker. Finally, the fabric is rolled up by a winding machine to prevent the formation of creases or tearing, as shown in Figure 4.

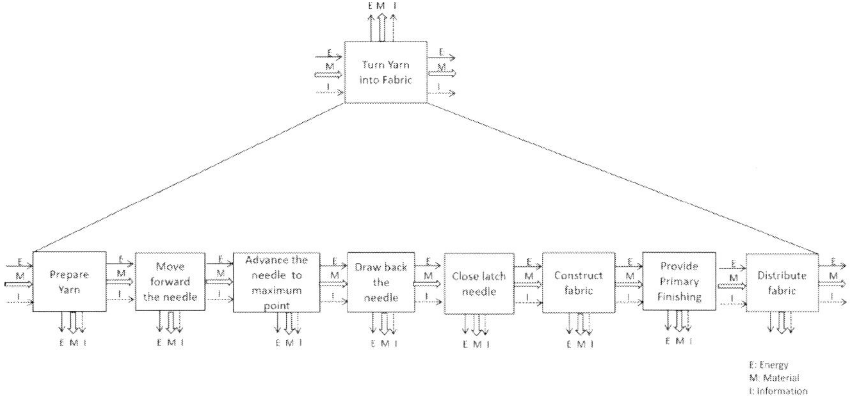

Figure 4. Turn yarn into fabric.

After the fabric has been made, it is necessary to prepare it for the finishing processes. The preparation for dyeing involves the impregnation of the fabric with a chemical solution for removal of impurities and paraffin from the yarn that hamper the absorption of dyes and pigments in the fibers using equipment known as the Over Flow, which is also responsible for the actual dyeing process. The softening of the fabric is held with the Foulard. The fabric is then dried and calendered. The distribution of the finished fabric is conducted through a transportation conduit and assessment system.

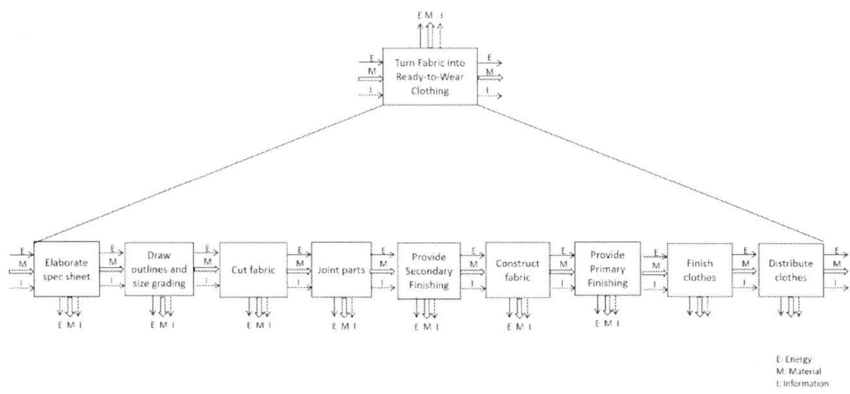

Figure 5. Turn fabric into ready-to-wear clothing (Authors).

The last productive chain system studied, "Turn Fabric into Ready-to-Wear Clothing," consists of several activities and components. This system includes the stages of development of the spec sheet, sampling, cutting, sewing and printing (Figure 5).

The spec sheet contains the technical drawings, fabric samples, measuring and sizes of the articles and notions used and can be done manually or with the aid of a software.

Sampling encompasses the drawing of outlines and size grading of the articles. The samples can be created manually or with the aid of a software.

The cutting of the fabric involves the planning of the cut outs and the cutting itself, which could be done manually or with the aid of computer programs and automatic machines. The planning of the cut outs requires knowing the characteristics of the fabric and the fitting of the sample and can be done manually or with the aid of a software. Finally, the cutting itself can be executed with the aid of electric tools such as saws or knives, or with an automatic machine. The articles cut are then sewn together with the use of sewing machines.

The dyeing of the fabric can be performed in the total (Dyeing) or partial (Printing) area of the textile substrate. Printing, specifically screen printing, requires screens and dryers. A print is engraved into the screen and decomposed according to the colors required for the formation of the image, and the preparation of the screen depends on the frame and form of fixation; the feeding of the machine is manual and the thermofixation is done with dryers. The finishing of the articles corresponds to the revision and quality inspection of the final product.

LCA Simulation

[31] reported an analysis of the life cycle of a knitwear, which produces Jersey, 100% cotton, carded, weight 330g/m^2. The authors concluded that cotton fiber production and processing are strongly related to soil impacts, acidification, and eutrophication. Knitting impacts radiation, inorganic

emissions, and climate change. Finally, the main impacts on the mesh processing process are ozone emission, organic emissions, and fossil fuels.

[32] conducted an operational and environmental research in a confection. In this research, the authors noted that 20% of all fabric used in the making and 30% of electrical energy was wasted. The authors do not present environmental impacts, as they focus only on economic aspects, specifically on production costs.

[33] and [34] concluded that cotton growing is responsible for most of the environmental impacts in the TC chain, due to the use of chemical agricultural inputs and land use for cultivation. Barbosa (2012) applied LCA to a spinning mill that analyzed yarn production and transportation. In turn, [34] analyzed the agricultural production of cotton for fiber production.

The application of LCA in the present research was based on data published by above authors, and were renamed to facilitate the simulation:

- System 1 (S1): corresponding to "Obtain Textile Fibers";
- System 2 (S2): corresponding to "Turn Fibers into Yarn";
- System 3 (S3): corresponding to "Turn Yarn into Fabric";
- System 4 (S4): corresponding to "Turn Fabric into Ready-to-Wear Clothing."

An overview is provided by a block diagram built by the openLCA program is shown in Figure 6.

The results of the Life Cycle Impact Assessment are summarized in Figure 7.

The acidification effect increases the acidity of the water and soil that forms "acid rain," while eutrophication promotes abnormal productivity of the elements of an ecosystem due to the concentration of chemical nutrients. Both impacts are related to decreased water quality and ecosystem biodiversity and are strongly related to fertilizer use [35]. The production of raw materials, in particular natural fibers such as cotton, is a major cause of acidification and eutrophication. However, it is noted that the systems "Turn Fibers into Yarns," "Turn Yarns into Fabrics" and "Turn Fabrics into Ready-to-Wear Clothing" present high values due to the chemicals used in the finishing process and the lack of effluent treatment, according to the results described by [36] and [37].

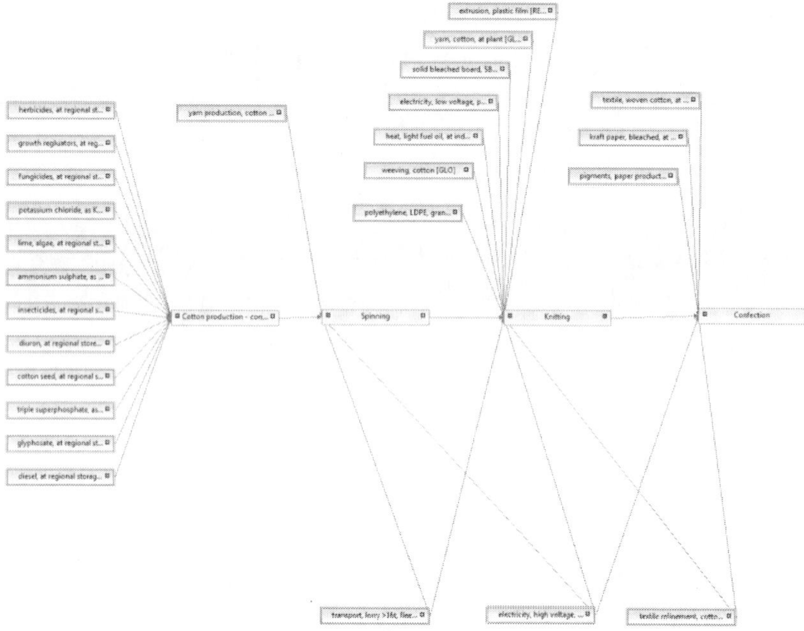

Figure 3. Block diagram of the production systems (Authors).

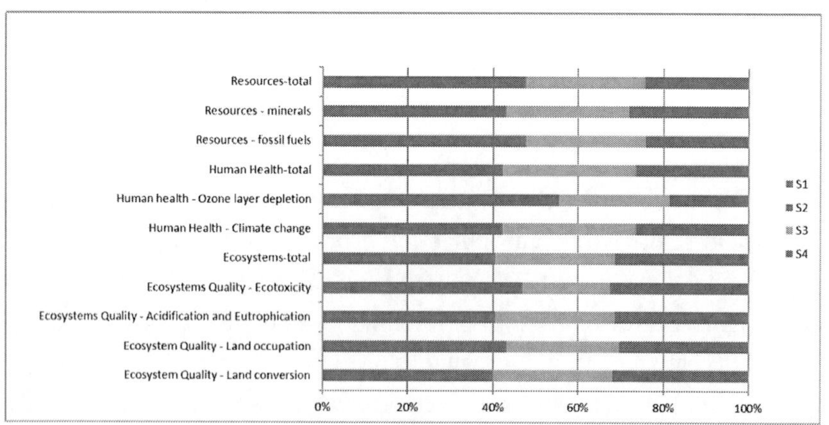

Figure 4. Simulation results (Authors).

According to [38], much has been done against the use of ozone-depleting substances, especially with global policies such as the Montreal Protocol (1987), London (1990) and Copenhagen (1992). The results indicate that all four processes contribute to the depletion of the ozone layer, which can have detrimental consequences for human health, natural ecosystems, marine life and animals due to increasing UVB rays entering the Earth's atmosphere [35]. [39] state that the textile industry has not shown significant technological development since the turn of the century in developing countries, which may be an argument for maintaining industrial practices incompatible with the contemporary unsustainable environmental reality.

Ecotoxicity is the impact of toxic substances on the ecosystem [23], and significantly influences biodiversity loss and species extinction. Its reference unit is expressed in unit, kg, 1,4-dichlorobenzene (1.4 DB) [35]. The results indicate that all processes have an ecotoxicity impact, mainly because they are related to organic and inorganic chemicals, pesticides, and fossil fuels.

Climate change is a change in global temperature caused by greenhouse gas emissions resulting mainly from industrial activities on a global scale. This impact is expressed as Global Warming Potential (GWP) in the reference unit, kg, CO_2 equivalent [35]. Climate change is observed in the "Turn Fibers into Yarns" and "Turn Yarns into Fabrics" systems, mainly because of the dyeing process, also described by [40].

Land conversion refers to a change in land cover and land occupancy refers to continuous use of an area [41]. The units expressed in these cases are the "potential species disappearance fraction" (PDF) times m2 and by PDF * m2 * year, respectively [35]. Both impacts analyze altered land area and soil and species loss in agriculture, forestry, mining, and transportation [42] which are observed in "Turn Fibers into Yarns," "Turn Yarns into Fabrics" and "Turn Fabrics into Ready-to-Wear Clothing."

The rate of energy consumption, including electricity and fossil fuels, is high in the textile industry. The amount of resource consumption corresponds to an unsustainable paradigm described by [43] as an industry contribution to environmental degradation and exploitation of nonrenewable resources.

Technologica Assessment and Forecasting

For the analysis of patents in the Assessment, the following project typologies were adopted:

- Original design: has a high degree of conceptual novelty and high degree of configuration complexity;
- Redesign: has a low degree of conceptual novelty and low degree of complexity in the changing of its configuration;
- Adaptive design: has a high degree of conceptual novelty and low degree of configuration complexity;
- Development design: has a low degree of conceptual novelty and high degree of complexity in the changing of its configuration.

The components of the production system "Obtain Textile Fibers" are classified as redesigns, since many patents described are adaptations of existing machines or improvements of the method. Examples of such adjustments include the various devices plugged into the soil preparation machines, harvester and feeder [44, 45, 46, 47, 48, 49] and the changes in the configuration of the machines [50, 51, 52]. The automatic devices that aid in the harvesting and manual ginning of cotton [53, 54] are considered development designs, because they assist in the execution of a function with complex variables without adding conceptual novelties. This also applies to the use of robots for the harvester [55] and the robotics platform for tillage [56]. Still in this system, the systems of communication, control and real-time assessment of the means of transport [57, 58, 59] are considered original designs, for having a high degree of complexity and conceptual novelty.

In the productive system "Turn Fibers into Yarns," most of the patents describe a redesign, when indicating the development of the constructive elements and of the methods of execution of the function with a low degree of conceptual novelty; examples of this design include the systems of opening of cotton bales [60], the additional carding device [61] and the development of the constructive elements of the rover, spinner [62] and foulard [63]. On the other hand, the quality control system of the Comber [64] is understood as a development design, due to its high degree of complexity and low degree of conceptual novelty.

The patents of the components of the production system "Turn Yarns into Fabrics" have development design [65, 66] and redesign [67, 68, 69] features. In addition, there are also examples of adaptive design, with conceptual changes in the component and a low degree of complexity, as in the case of the needle and sinker patents that prevent the accumulation of cotton fibrils [70, 71].

Finally, in the productive system "Turn Fabrics into Ready-to-Wear Clothing," the components are classified as variant design and development design. The patents of the variant design components are adaptations of the elements of the screen for printing, dryers [72, 73], cutting table [74], sewing machine [75, 76] and iron [77, 78]. The development design components are related to the control system of positioning of the screen for printing [79], and to the positioning of the fabric on the sewing machine [80]. The automatic cutting table is the component associated with the development design, bearing high complexity and low degree of conceptual novelty.

In general terms, it is possible to notice that there are few original designs among the components described; the patents examined indicate adaptations to the machines or addition of components to the control system of the equipment. Thus, the importance of technology forecasting in the determining of technologies that have already been developed, though not yet applied in their entirety, and in the identification of possible new technologies that meet the demands of the Fourth Industrial Revolution, is notable.

Technological Forecasting indicates the need for technological change in all production systems. For the obtaining of textile fibers, agricultural implements must obtain and manage information in real time pertaining to the weather and soil conditions, to the quality of the seeds and inputs to assist the process of decision-making associated with the actions involved in the processes from cultivation until harvest. There are patents that concern the systems of data acquisition and management [81], the transmission of data via the IoT to the consumer [82] and the management of climatic conditions [83], but these tools are not extensively used in agricultural machinery yet.

The Industry 4.0 will change the organization of factories and business models. Two models are widely used: the intelligent factory and the mini-factory. The Intelligent Factory has as principle the integration between various stakeholders in a virtual and collaborative environment, the remote control and assessment of industrial activities through virtual copies of the physical environment and the intensive use of ubiquitous technologies in portable devices.

Mini-factories are based on on-demand production, in which the production depends directly on consumer demand. To produce yarns and fabrics, the remote assessment of machines and environmental conditions and the use of a QR Code (Quick Response Code) and RFID (Radio Frequency IDentification) for data access, are examples of additional machinery components. It is worth mentioning that the information contained in the monitoring devices are in packaging and labels that will be discarded in the next steps of textile production. Thus, the information needs to be contained on the yarn.

The confection stage can be reconfigured in a mini-factory, which encompasses the orders, sampling, dyeing, cutting, sewing, finishing and shipping departments. It is worth noting that in this mini-factory, all components must be automated, an original design being needed for each machine.

CONCLUSION

Given the present moment of constant modification, connectivity, digitization of industrial processes, interaction between people and machines, decentralization of decision making and the need to adopt sustainable practices, it is noted that the development process and life cycle analysis of a product will form a basis for understanding the influence of the use of new technologies on the production process. It was possible to develop a reference model that contemplated the functional, technological and environmental aspects of t-shirt production to describe the current scenario of some textile production systems and to make a technological prediction of the components given the context of the Fourth Industrial Revolution. The chaining of subfunctions, the specifications of the input and output flows of material, information and energy and the interrelationship with the components allowed a clear view on the productive processes of the TC Chain and the interdependence of all actors. LCA corroborates the importance of analyzing environmental impacts along the production chain; Among the systems analyzed in this research, it is noted that all stages play a part in generating negative impacts on human health, climatic, atmospheric and soil conditions, but the system for transforming fibers into yarns has the worst environmental indexes.

REFERENCES

[1] Ha-Brookshire, J., LaBat, K. 2015. "Envisioning textile and apparel research and education for the 21st century." International Textile and Apparel Association – ITAA Monography.

[2] Ülgen, V. S., Forslund, H. 2015. "Logistics performance man-agement in textiles supply chains: best-practice and barriers." *International Journal of Productivity and Performance Management*, 64 (1): 52 – 75. doi: 10.1108/IJPPM-01-2013-0019.

[3] Lu, S. C.-Y, Elmarachy, W., Schuh, G., Wilhelm, R. 2007. "A Scientific Foundation of Collaborative Engineering." *Annals of the CIRP*, 56/2/2007:605-634. Doi: 10.1016/j.cirp.2007.10.010.

[4] Mcneil, Ian. 1990. *An Encyclopedia of the History of Technology*. 1st Edition, London, Routledge, 1081 p.

[5] Flacher, D. 2005. "Industrial Revolutions and Consumption: A Common Model to the Various Periods of Industrialization." Working paper, halshs-00132241.

[6] Anderl, R. 2015. "Industrie 4.0 – technological approaches, use cases, and implementation." *at-Automatisierungstech.*, 63(10): 753-765. doi: 10.1515/auto-2015-0025.

[7] Schuh, G., Potente, T., Varandani, R., Hausberg, C., Fränken, B. 2014. "Collaboration moves productivity to the next level." Procedia CIRP 17, p.3-8. Doi: 10.1016/j.procir.2014.02.037.

[8] Hermann, M., Pentek, T., Otto, B. 2015. "Design principles for Industrie 4.0 Scenarios: a literature review." Working Paper n.01/2015, Technische Universität Dortmund.

[9] Gorecky, D., Schmitt, M., Loskyll, M., Zuhlke, D. 2014. "Human-Machine-Interaction in the Industry 4.0 Era." In: Industrial Informatics (INDIN), 12th IEEE International Conference, p. 289-294. Doi: 10.1109/INDIN.2014.6945523.

[10] Castells, Manuel. *The rise of the network society: The information age: Economy, society, and culture*. John Wiley & Sons, 2010, ISBN 978-1-4051-9686-4.

[11] Gölzer, P., Cato, P., Amberg, M. 2015. "Data Processing Requirements of Industry 4.0-Use Cases for Big Data Applications." ECIS 2015 Proceedings, 2015.

[12] Kagermann, H., Wahlster, W., Helbig, J. 2013. "Recommendations for implementing the strategic initiative Industrie 4.0: final report of the Industrie 4.0 Working Group." 82p.

[13] Cheng, S. 2015. "Why the Fourth Industrial Revolution needs to be global". Acessed December 2015. http://www.weforum.org/agenda/2015/11/why-the-fourth-industrial-revolution-needs-to-be-global.

[14] Honoré, Carl. 2009. *In praise of slowness*. Harper Collins, ISBN 978-0-06-1907296.

[15] Sabo, F. 2015. "Industry 4.0 – a comparison of the status in Europe and the USA." Austrian Maschall Plan Foundation, 33p.

[16] Židonienė, S., Kruopienė, J. 2015. "Life Cycle Assessment in environmental impact assessments of industrial projects: towards the improvement." *Journal of Cleaner Production*, 106: 533-540. Doi: 10.1016/j.jclepro.2014.07.081.

[17] Parisi, M. L., Fatarella, E., Spinelli, D., Pogni, R., Basosi, R. 2015. “Environmental impact assessment of an eco-efficient production for coloured textiles.” *Journal of Cleaner Production*, 108: 514-524. doi: 10.1016/j.jclepro.2015.06.032.

[18] Brundtland, Gro Harlem. 1987. *World commission on environment and development: Our Common Future*. Oxford: Oxford University Press.

[19] Tobler-Rohr, Marion I. 2011. *Handbook of sustainable textile production*. Ed: Woodhead Publishing, ISBN 9780857091369.

[20] Crul, M. R. M.; Diehl, J. C; Ryan, C. 2009. "Design for sustainability: a step-by-step approach." UNEP, 110p.

[21] ISO 14040: 2006. *Environmental management – Life Cycle Assessment – Principle and framework*, 20p.

[22] Ramani, K., Ramanujan, D., Bernstein, W. Z., Zhao, F., Sutherland, J., Handwerker, C., Choi, J., Kim, H., Thurston, D. 2010. “Integrated sustainable life cycle design: a review.” *Journal of Mechanical Design*, 132 (9): 091004/1-091004/15. doi: 10.1115/1.4002308.

[23] Guinée, Jeroen B. 2002. *Handbook on Life Cycle Assessment: operational guide to the ISO standards*. ISBN: 0-306-48055-7, v.7, 687p.

[24] Hellweg, S., I-Canals, L. M. 2014. “Emerging approaches, challenges and opportunities in life cycle assessment.” *Science*, 344(6188): 1109-1113, 2014. doi: 10.1126/science.1248361.

[25] Allwood, J. M., Laursen, S. E., de Rodriguez, C. M., Bocken, N. M. P. 2006. *Well dressed? The present and future sustainability of clothing and textiles in the United Kingdom*. Univesity of Cambridge, ISBN 1-902546-52-0, 84p.

[26] ABIT – Brazilian Association pf Textile and Apparel Industry. 2014. “Agenda – 2015/2018.” São Paulo, Brazil, 19p. (in Portuguese).

[27] González, J. A. 2013. “La sostenibilidad ecológica en el desarrollo de productos textiles: Una Revisión de Literatura.” *Realidad y Reflexión*, 38: 65-97, doi: doi.org/10.5377/ryr.v38i0.1833. (in Spanish)

[28] Eckert, Claudia. 2015. “Design for Values in the Fashion Fashion and Textile Textile Industry.” In *Handbook of Ethics, Values, and*

Technological Design: Sources, Theory, Values and Application Domains, edited by van den Hoven, Jeroen, Vermaas, Pieter E., van de Poel, Ibo, 691-715. Springer Science+Business Media Dordrecht.

[29] Gloy, Y., Schwarz, A., Thomas, G. 2013. "Cyber-physical systems in textile production: the next industrial revolution?" *Proceedings of the 1st International Conference on Digital Technologies for the Textile Industry*, 5-6 September, Manchester – UK.

[30] Pahl, G.; Beitz, W.,; Feldhusen J.; Grote K. H. 2007. *Engineering Design: a Systematic Approach.* 3 Ed. Springer-Verlag London Limited.

[31] Santos, A. P. L., Fernandes, D. S. 2012. "Análise do impacto ambiental gerados no ciclo de vida de um tecido de malha." *Iberoamerican Journal of Industrial Engineering*, 4(7): 1-17. (in Portuguese).

[32] Pimenta, H. C. D., Gouvinhas, R. P. 2011. "Implementação da produção mais limpa na indústria têxtil: vantagens econômicas e ambientais." *3rd International Workshop: Advances in Cleaner Production,* 18-20 May, São Paulo, Brazil. (in Portuguese).

[33] Barbosa, Priscila Pasti. 2012. "Análise dos impactos ambientais da cadeia têxtil e do algodao no espaço urbano-industrial: uma aplicação da avaliação do ciclo de vida." Master Thesis, State University of Maringa, Brazil. (in Portuguese).

[34] Donke, A. C. G., Tereza, L. D. S. B. M., Dias, S. N., de Figueirêdo, S. M. C. B., da Silveira, L. K. M. I., Matsuura, F. 2013. "Life cycle impact assessment of cotton production in the Brazilian Savanna." *Proceeding of the Vth International Conference on Life Cycle Assessment - CIACV*, p. 189-195.

[35] Acero, A. P., Rodríguez, C., Ciroth, A. 2015. "LCIA methods: impact assessment methods in Life Cycle Assessment and their impact categories." Greendelta, version 1.5.4, 23p.

[36] Terinte, N., Manda, B. M. K., Taylor, J., Schuster, K. C., Patel, M. K. 2014. "Environmental assessment of coloured fabrics and opportunities for value creation: spin-dyeing versus conventional dyeing of modal fabrics." *Journal of Cleaner Production*, 72: 127-138. doi: 10.1016/j.jclepro.2014.02.002.

[37] Yuan, Z. W., Zhu, Y. N., Shi, J. K., Liu, X., Huang, L. 2012. "Life-cycle assessment of continuous pad-dyeing technology for cotton fabrics." *International Journal of Life Cycle Assessment*, 18 (3), 659e672. doi: 10.1007/s11367-012-0470-3.

[38] Middlebrook, Ann M., Tolbert, Margaret A. 2000. *Stratospheric Ozone Depletion*. University Science Book: USA, ISBN 1-891389-10-6.

[39] Bhaumik, S., Dimova, R. 2014. "Good and bad institutions: is the debate over? Cross-country firm-level evidence from the textile industry." *Cambridge Journal of Economics*, 38(1): 109-126.

[40] Zhang, Y., Liu, X., Xiao, R., Yuan, Z. 2015. "Life cycle assessment of cotton shirts in China." *International Journal of Life Cycle Assesment*, 20(7): 994-1004. doi: 10.1007/s11367-015-0889-4.

[41] Mattila, T., Helin, T., Antikainen, R., Soimakallio, S., Pingoud, K., Wessman, H. 2011. *Land use in life cycle assessment*. The Finish Environment Institute, ISBN 978-952-11-3926-0.

[42] Finnveden, G., Hauschild, M.Z., Ekvall, T., Guinée, J., Heijungs, R., Hellweg, S., Koehler, A., Pennington, D., Suh, S. 2009. "Recent developments in life cycle assessment." *Journal of Environmental Management*, 91(1): 1-21.

[43] Wang, S., Wan, J., Li, D., Zhang, C. 2015. "Implementing Smart Factory of Industry 4.0: an outlook." *International Journal of Distributed Sensors Network,* P. 1-11, 2015. Doi: 10.1155/2016/3159805.

[44] Orlanda, F. J. Q. 2014. 2014. Dispositivo Para Aumento Da Faixa De Preparo De Solo. BR 10 2012 004364 5 A2. (Patent in Portuguese).

[45] Rylander, D. J. 2013. Ground working machine with a blockage clearing system and method of operation. U.S. Patent n. 8,408,149, 2 April 2013, 5 p.

[46] Silva, M. N.; Silva, J. J. C; Bortolan, M. T.; Calegaro, J. C.; Kangerski, A. L.; Idehara, S. J. 2015. Implemento Agrícola De Preparo De Solo Para Conformação De Camalhões. BR 10 2013 027417 8 A2, 01 set. 2015. (Patent in Portuguese).

[47] Marchesan, J. C. 2016. Disco Agrícola Para Implementos Agrícolas Para Preparo Do Solo. BR 10 2014 025650 4 A2, 17 maio 2016. (Patent in Portuguese).

[48] Klein, M. R. 2014. Levantador De Culturas Aplicável Em Uma Plataforma De Corte De Uma Máquina Colhedora. BR 10 2012 026731 4 A2, 30 set. 2014. (Patent in Portuguese).

[49] Bertino, L. H. 2013. Implemento Ceifador Enleirador para Culturas Diversas. BR Patent, BR 10 2012 025615 0 A2, 15 fev. 2013. (Patent in Portuguese).

[50] Henkels, E. 2015. Processo De Coleta E Prensagem De Resíduos De Algodão. BR 10 2014 005897 4 A2, 01 dez. 2015. (Patent in Portuguese)

[51] Seki, A. S.; Balestrin, L. W. A. 2017. Plataforma Para Uma Máquina De Colher. BR 10 2015 015922 6 A2, 02 fev. 2017. (Patent in Portuguese).

[52] Yang, Y; Yuancheng, Y; Yan, G.; Li, L. X.; Zheng, H.; Jiang, H.; Chen, Y.; Chen, Y.; Liu, X.; Zhang, S. 2013. Short fiber packer. CN203046287 U, 10 jul. 2013, 9 p.

[53] Xiuyun, L. 2012. *Micro hand-operated cotton gin.* CN202925156 U, 8 may 2013, 5 p.

[54] Jun, Z.; Liang, X. 2016. Handheld Automatic Cotton Picker. CN104871731 B, 16 nov. 2016, 5 p.

[55] Kahani, A. 2016. Multi-robot crop harvesting machine. U.S. Patent n. 9,475,189, 25 out. 2016, 22 p.

[56] Cavender-Bares, K.; Bares, C. C. 2015. Robotic platform and method for performing multiple functions in agricultural systems. U.S. Patent Application n. 15/047,076, 28 may 2015.

[57] Foster, C. A.; Ringwald, J. R.; Fitzkee, Douglas S.; Richman, K. S.; Orsborn, J. H.; Brown, T. H.; Posselius, J. H.. 2013. System and method for automatically updating estimated yield values. U.S. Patent Application n. 13/903,624. 18 may 2013, 20 p.

[58] Ferguson, D. I..; Dolgov, D. A. 2013. Modifying behavior of autonomous vehicle based on predicted behavior of other vehicles. U.S. Patent n. 8,457,827, 4 jun. 2013, 18p.

[59] Santos, R. L. 2016. Sistema Para Controle Monitoramento E Coleta De Dados De Atuadores Embarcados Em Caminhão. BR 10 2015 031811 1 A2, 29 nov. 2016. (Patent in Portuguese).

[60] Huazhong, L. 2014. Improved fully-automatic unstacking and unpacking system. CN203667100 U, 25 jun. 2014, 7 p.

[61] Muller, U. 2015. Elemento De Cardagem E Carda Circular De Uma Máquina De Cardagem E Máquina De Cardagem. BR 10 2013 006467 0 A2, 09 jun. 2015. (Patent in Portuguese).

[62] Havliczek, J. 2013. Engrenagem Trocadora De Guia-Fios Para Filatório De Dupla Face Sem Fuso. PI 1004878-2 A2, 12 março 2013. (Patent in Portuguese).

[63] Zhao, M. 2013. Foulard. CN Patent, CN205062446U, 02 aug. 2013, 5 p.

[64] Zhi, R. 2012. Device for testing noil rate of combing machine. CN Patent, CN202415781 U, 5 sept. 2012, 8 p.

[65] Min, Z.; Zhao, Q.; Zhou, X. 2013. Non-winding take-up device for warp knitting machine. CN203411153U, 29 jan. 2014, 9 p.

[66] Aramaki, Y.; Shimosakoda, K. 2013. Method of and device for controlling fabric take-up in electronic pattern knitting machine. EU Patent EP2546401 A1, 16 jan. 2013, 19p.

[67] Chen, M. C. 2015. Mecanismo De Alimentação De Máquina De Costura. BR Patent, BR 10 2014 013621 5 A2, 13 outubro 2015, 20 p. (Patent in Portuguese).

[68] Gao, M. Z. 2013. Adjustable needle base. CN Patent, CN202989394 U, 12 jun. 2013, 6 p.

[69] Mayer, K. T. G. 2012. Yarn guide arrangement of a warp knitting machine. DE Patent, DE202012005478U1, 26 jun. 2012, 5 p.

[70] Hong, R. 2016 Knitting machine needle construction prevents the accumulation of cotton. CN205501563U, 26 jun. 2016, 9 p.

[71] Tian, J-M. 2013. Dustproof latch needle. CN Patent, CN102535000 B, 11 dec. 2013, 5 p.

[72] Zonggen, Z. 2012. Full-automatic double-faced silk printing drying all-in-one machine. CN Patent, CN202428761U, 12 sept. 2012, 8 p.

[73] Hwa, K. J. 2016. Silk screen plate dryer, KR Patent, KR101661754B1, 6 nov. 2016, 11 p.

[74] Tong, Z. S. 2012. Textile cutting workbench and cutting machine adopting same CN202152429U, 29 feb. 2012, 21 p.

[75] Chen, M. C. 2015. Mecanismo De Alimentação De Máquina De Costura. BR Patent, BR 10 2014 013621 5 A2, 13 outubro 2015, 20 p. (Paten in Portuguese).

[76] Medeiros, L. R. R. 2015. Acessório Para Máquina De Costura. BR 20 2014 012064 0 U2, 15 dez. 2105. (Patent in Portuguese).

[77] Ong, C. K. 2016. Dispositivo De Passar Roupa A Vapor. BR 11 2013 015661 9 A2, 11 out. 2016. (Patent in Portuguese).

[78] Soares, E. B. F. 2013. Ferro De Passar Roupa Com Compartimento Para Ajudar Com Amaciante. BR Patente, MU 9101887-0 U2, 13 agosto 2013. (Patent in Portuguese).

[79] Xingxing, H. 2012. Positioning method in novel silk screen printing CCD (charge coupled device) image identification. CN Patent, CN102328493A, 25 jan. 2012, 10 p.

[80] Guerreschi, C. 2016. Aparelho Para O Posicionamento De Peças De Tecido Sobre Máquinas De Costura. BR 10 2016 005873 2 A2, 20 set. 2016. (Patent in Portuguese).

[81] Curkendall, L. D.; Pape, W. R.; Dolan, A. J. 2006. Method and system for agricultural data collection and management. U.S. Patent n. 6,995,675, 7 feb. 2006, 85p.

[82] Dlott, J. W.; Rosenberg, H. W.; Ohmart, C. P.; Connolly, R. 2008. Method and system to communicate agricultural product information to a consumer. U.S. Patent n. 7,440,901, 21 oct. 2008, 38 p.

[83] Frey, M. M. 2014. Systems, devices, and methods for environmental monitoring in agriculture. U.S. Patent Application n. 14/017,182, 19 jun. 2014, 26 p.

Biographical Sketches

Adriana Yumi Sato Duarte, PhD

Affiliation: Nossa Senhora do Patrocínio University

Education: PhD, Mechanical Engineering

Business Address: Praça Antônio Vieira Tavares, Centro, Salto, SP –Brazil

Research and Professional Experience: Undergraduate (2009) in Bachelor of Textiles and Fashion from the University of São Paulo, Master (2013) and PhD (2017) in Mechanical Engineering from the State University of Campinas (Unicamp). Conducted a period of Internship of Doctorate Sandwich Abroad (SWE) -Science without Borders Program (2015-2016) at Fachgebiet Datenverarbeitung in der Konstruktion (Dik), Technical University of Darmstadt, Germany. She has experience in Mechanical Engineering with an emphasis on Mechanical Design and in Textiles and Fashion with an emphasis on product design methodology, sustainable product development, Brazilian natural fibers, knitting technology and Industry 4.0. She is currently Assistant Professor II at Nossa Senhora do Patrocinio University and Coordinator of the Fashion Design Course.

Publications from the Last 3 Years: 15

Regina Aparecida Sanches, PhD

Affiliation: University of Sao Paulo Education: Postdoctorate in Design

Business Address: Avenida Arlindo Bettio, 1.000, Sao Paulo-SP – Brazil

Research and Professional Experience: Degree in Textile Engineering at University Center of FEI (1987), Master in Mechanical Engineering at State University of Campinas (2001), PhD in Mechanical Engineering at State University of Campinas (2006) and Postdoctorate in Design at University of Lisbon (2016). She started her academic career in 1995, was the coordinator of the undergraduation course in Textile Engineering at University Center of FEI (2001 to 2006), was the coordinator of the undergraduation course in Textile and Fashion at University of Sao Paulo (2010 to 2012), was the coordinator of the Master's Degree in Textile and Fashion at University of Sao Paulo (2012 to 2016). She has been a professor at the School of Arts, Sciences and Humanities since 2006 and has been an associate professor at the University of São Paulo since 2011. She's a visiting professor at University of Campania Luigi Vanvitelli (Italy), at the University of Lisbon (Portugal) and at the Polytechnic Institute of Castelo Branco (Portugal) and researcher at the Center for Research in Architecture, Urbanism and Design (CIAUD) at the University of Lisbon (Portugal). She researches in the areas of textile materials, knitting technology and textile design.

Publications from the Last 3 Years: 82

Rayana Santiago de Queiroz

Affiliation: Institute for Technological Research

Education: PhD student

Business Address: Avenida Professor Almeida Prado, 532, Sao Paulo-SP – Brazil

Research and Professional Experience: PhD student in the Textile Engineering course at the University of Minho (Portugal), master (2013) and graduated (2009) by the Textile and Fashion course at the University of São Paulo. Since 2012 acts as a researcher at the Technical Textiles and

Protection Products Laboratory of the Institute for Technological Research, where has been working especially on the following topics: vegetable textile fibers, natural dyes, comfort, characterization and performance evaluation of technical textiles.

Publications from the Last 3 Years: 11

Fernando Soares de Lima

Affiliation: Institute for Technological Research

Education: Master in Industrial Processes

Business Address: Avenida Professor Almeida Prado, 532, Sao Paulo-SP – Brazil

Research and Professional Experience: Degree in chemistry from the University of Mogi das Cruzes (2004), Master in Industrial Processes from the Technological Research Institute of the State of São Paulo (2013) and Chemical Production Engineer from Faculdades Oswaldo Cruz (2017). He is currently responsible for the Technical Textiles and Protective Products Laboratory and for the Shoes and Protective Products Laboratory of the Technological Research Institute of the State of São Paulo. He mainly works on the following topics: technical fabrics, characterization tests and performance evaluation of textiles and PPE's, weathering and microencapsulation applied to textiles.

Publications from the Last 3 Years: 7

Franco Giuseppe Dedini, PhD

Affiliation: State University of Campinas (Unicamp)

Education: PhD Mechanical Engineering

Business Address: Mendeleyev St, 200 - Cidade Universitária, Campinas - SP, Brazil

Professional Experience: Undergrate in Mechanical Engineering from the State University of Campinas (1980), Master's degree in Mechanical Engineering from the State University of Campinas (1985) and a PhD in Applied Mechanics from the Polytechnic of Milan (1993). He is currently an associate professor - MS5 at the State University of Campinas, Reviewer of Product (1676-4056), *International Journal of Quality and Reliability Management, SAE Technical Papers, Brazilian Journal of Mechanical Sciences and Science & Engineering*. He has experience in Mechanical Engineering, with emphasis on Mechanical Design mainly on the following themes: product development, vehicle dynamics, machine design and design methodology.

Publications from the Last 3 Years: 30.

In: Challenges and Opportunities …
Editor: Wallace G. Tarrant
ISBN: 978-1-53618-770-0

Chapter 5

THE USAGE OF IMAGE PROCESSING TECHNIQUES ON THE DETERMINATION OF PILLING GRADES

Abdurrahman Telli*

Department of Textile Engineering, Cukurova University, Adana, Turkey

ABSTRACT

Pilling is a serious defect of fabric surface that gives an unpleasant appearance to garment. Pilling tendency is tested with different methods and devices in the laboratory conditions. The determination of the pilling grades is made with visual control by operators. Therefore, the human factor is significantly effective in this subjective evaluation method and may cause incorrect results. Studies in recent years show that objective methods based on image processing are preparing to replace subjective pilling assessments. In this chapter, difficulties in the subjective evaluation of the pilling grades were explained. Potential opportunities presented by image processing studies in the literature on the objective evaluation of the pilling grades were investigated. Image processing steps were given with

* Corresponding Author's Email: atelli@cu.edu.tr.

various examples by using Image Processing Toolbox and codes in MATLAB software. In this study, it was indicated that it is possible to make an objective pilling detection easily for the fabric structures used as standard in the textile industry thanks to the databases to be created with measuring lots of samples.

Keywords: pilling grades, pilling tendency, pilling resistance, image processing, MATLAB

INTRODUCTION

Pilling is a serious defect of fabric surface that gives an unpleasant appearance to garment. Pills can be formed during use or washing. It is the result of the rubbing movements of the fabric with its own fabric or on different surfaces. Figure 1 shows two different knitted fabrics consisted of different pill structures.

Figure 1. Two different knitted fabrics consisted of different pill structures.

A remarkable amount of literature has been focused on the definition of pilling, pill mechanism, mechanism of pill formation, different techniques in determination of pilling grades, parameters causing pill formation and several physical or chemical methods for decreasing pilling. There is a large volume of published previous studies describing the role of production parameters on pilling. The pilling performance of fabrics is influenced by many production parameters such as material type (fiber type, fiber fineness, fiber length and length distribution, fiber strength and bending rigidity, fiber cross-sectional shape, inter-fiber relationship and fiber crimp), yarn technology (yarn spinning system, yarn twist, yarn count, yarn hairiness, blending ratio and yarn plying), fabric production technology (fabric type, fabric density, fabric weight, construction type), dyeing & finishing technology (physical or chemical finishing processes) [1-5]. It can be seen in earlier studies that so many parameters on pilling grade have been identified. However, the reported data are rather controversial about synergistic effect of each on pilling degree and there is no general agreement about effects on pilling. When findings to emerge from previous studies are taken as a whole, we can surmise that each process step may weaken the effect of the previous operation or remove it completely with right a choice. The pills formed on the fabric surface consist of fibers. These fibers come from the yarn structure. Any combination that keeps the fibers more firmly in the yarn structure will increase the pilling grade. On the contrary, any change that allows fibers to migrate to the yarn surface will decrease the pilling degree [6]. In general, previous findings indicate that fabric pilling is principally connected to the fabric structure. Any combination that keeps the yarns more firmly in the fabric structure will increase the pilling grade. It can be said as the most important point that fiber movements should be blocked in yarn and fabric structure without harming fabric hand feeling for the more durable and better pilling performance of garments [7].

Test results of pilling resistance play an essential role in determining the quality and economic life of the garment. Pilling tendency is tested with different methods and devices in the laboratory conditions. In these methods, the use conditions of fabrics in daily life are simulated. Today, the most preferred pilling testers work with the Martindale test method. The Martindale device is used both for measuring pilling resistance and for measuring the abrasion resistance of fabrics. Martindale Pilling Tester and equipments are presented in Figure 2.

After the experiment is carried out following the desired standard, the stage of grading of the pilling is started. The determination of the pilling grades is made with visual control by operators. This rating according to EN ISO 12945-2 standard is presented in Table 1.

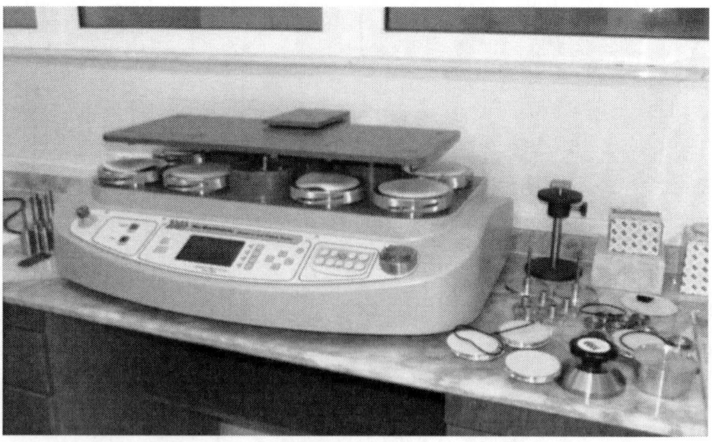

Figure 2. Martindale Pilling and Abrasion Tester.

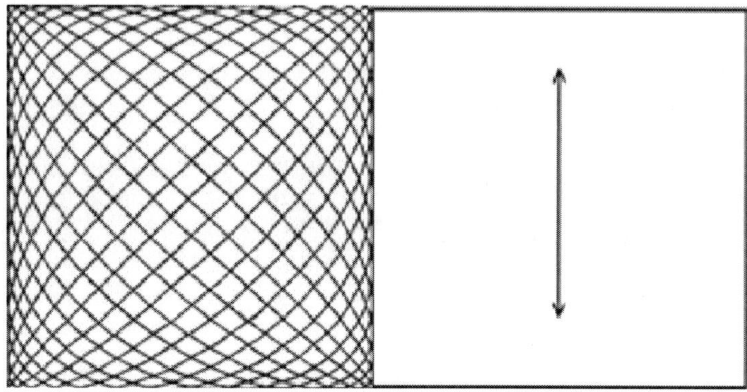

Figure 3. Lissajous pattern (left) and linear motion (right).

Table 1. Pilling Grading [8]

Grade	Description
5	No change
4	Slight surface fuzzing and/or partially formed pills
3	Moderate Surface fuzzing and/or moderate pilling. Pills of varying size and density partially covering the specimen surface
2	Distinct surface fuzzing and/or distinct pilling. Pills of varying size and density covering a large proportion of the specimen surface
1	Dense surface fuzzing and/or severe pilling. Pills of varying size and density covering the whole of the specimen surface.

The pilling event is carried out by the rubbing of the fabric on its surface with the Lissajous movement. Lissajous Pattern is presented in Figure 3.

Apart from the grades given in Table 1, intermediate ratings such as 1-2, 2-3, 3-4, 4-5 can be also made. Furthermore, this assessment is supported by standard photographs. EMPA standard photographs have been developed for the evaluation of samples studied on the Martindale pilling tester according to EN ISO 12945-2. The EMPA (Saint Gall, Switzerland) standard includes 24 photos in total. Half of these photographs are prepared for woven and the other half for knitted fabrics. There are three separate categories, each consisting of four images. The photographs in two separate groups for knitting (K1, K2, K3) and woven fabrics (W1, W2, W3) have three categories (low, medium and high yarn density) that allow the closest reference to be selected in the evaluation of samples. In EMPA images, the structure of the fabric background is often seen in high contrast. The pills can be distinguished by the regional changes they create, unlike the usual structure of the background. Since the lighting is in a horizontal direction, the pills appear in light color next to the shadows formed in the background [3].

Considering the aforementioned as a whole, it can be seen that there are standardized objective methods in the pilling of fabrics under certain conditions. However, evaluations of the pilling ratings in the next step are open to discussion because current methods are subjective methods. Even among experts, differences of opinion may arise. Although there are theoretical studies and an improved test device for objectively evaluating pilling, it is still a common method to compare pilling samples with standard photographs. Today, there is no practical and objective method that does not change according to the individual. The human factor is significantly effective in this subjective evaluation method and may cause incorrect results.

THE DIFFICULTIES IN THE CONVENTIONAL EVALUATION OF THE PILLING GRADES

Firstly, experienced staff is needed for an accurate assessment. However, obtaining the same results for piling grades is difficult because of visual perception differences even among experienced staff because there are various factors in the formation of pilling. The patterns on the fabric or different fabric and yarn structures can prevent the detecting of pills. In addition to these, visual control can be boring when done for a long time and defects can be overlooked. Each of these is misleading factors for the evaluator. The differences of opinion amongst evaluators cause serious economic and commercial problems in some cases.

Table 2. Test results and same grades [9]

Fabric Code	2000			5000		
	Author	Private Institution	Governmental institution	Author	Private Institution	Governmental institution
	I	II	III	I	II	III
1	4	3/4	2/3	3 (*)	3 (*)	2
2	5	4	3/4	4/5	3/4	3
3	2/3	3/4	2	1/2 (+)	3	1/2 (+)
4	3 (*)	3 (*)	2/3	2 (+)	2/3	2 (+)
5	3 (*)	3 (*)	2	2/3	3	1/2
6	4	3/4	2/3	3 (*)	3 (*)	2
7	3/4	3	2/3	2/3	3	2
8	4/5	4	3/4	4 (*)	4 (*)	3
9	3/4	3	2/3	2/3 (*)	2/3 (*)	2
10	4	3/4	3	3 (*)	3 (*)	2
11	3/4	4	3	2/3 (+)	3/4	2/3 (+)
12	4/5	3/4	3	4	3/4	2/3

(*) The same grade between I and II groups.

(+) The same grade between I and III groups.

(x) The same grade between II and III groups.

A research by Telli (2019a) was focused on human factors in the determination of pilling degree of knitted fabrics as subjective [9]. In the study, Nu-Martindale Test Instrument was chosen to provide pilling formation in the fabrics. The tests were carried out at 2000 and 5000 turns. The determination of pilling grading was based on Table 2 in compliance with EN ISO 12945-2. Moreover, this assessment was supported using EMPA Knitted standard photos. The produced twelve different knitted fabrics were sent to two different organizations for evaluation of pilling degrees. These were selected from accredited institutions by TURKAK for pilling testing. One of them was chosen from a governmental institution. Another organization was a private institution. The pilling degree results obtained from three different departments were statistically evaluated using correlation and reliability analysis on the SPSS program. The results obtained from the visual evaluation of knitted fabrics in their study are shown in Table 2. Table 2 compares the results obtained from three different groups.

It can be seen from the data in Table 2 that there were significant differences between the three groups. The same grades were not given for any samples by the three groups. There were seven same grades (29,17%) between "I" and "II". Two same grades (8,3%) was found between "I" and "III". Therefore, reliability analysis between all groups was tested. The researcher indicated that there was no problem caused by the experiment set or method according to the results of reliability analysis. After that, the decrease of sensitivity with 0.5 unit differences has provided a significant increase in similarities. In this case, eighteen similar grades (75%) were observed between "I" and "II". There were eight similar grades (33,3%) between "I" and "III". When all the item statistics and the results obtained from Pearson correlation analysis can be compared, more similar results among "I" and "II" were found in comparisons between the three groups. Moreover, the highest correlation amongst groups was seen between "I" and "III" (r = 0,873**). Researchers explained this situation as follows:

1. The visual perception of "III" was more negative than I although there was a strong correlation between the two groups. Because, the

lowest mean values were determined by "III". The lowest correlation was measured between I and II although the highest similar and same results were seen among "I" and "II".

2. It seems possible that their some results showed different tendency according to others. The first group have higher values than the second group in three of six dissimilar results. In the others, the second group received higher values. The trend is not in a fixed direction. However, the third group showed lower values than both groups except for the three values that are equal to the first group.

In general, it can be said that there is strong correlation between experts in the determination of the pilling grades with the conventional evaluation methods. However, differences in visual perception between experts make it difficult to obtain the same results. As can be easily seen from the example above, the same grades were not given for any samples. Similar findings were provided with the decrease of sensitivity with 0.5 units. In subjective evaluation, the human factor was clearly seen.

However, differences in 0.5 units can be tolerated at high scores such as 5, 4/5 or 4. Serious problems may occur at the values below, depending on the desired quality value. The different evaluations in different test laboratories may lead to conflicts resulting in high costs in the supply chain. Therefore, further works should be intensified on objective methods such as image processing in the evaluation of pilling.

Image Processing Studies as Potential Opportunities on the Objective Evaluation of the Pilling Grades

Studies in recent years show that objective methods based on image processing are preparing to replace subjective pilling assessments. Basically, in an objective evaluation, the image or video recording of the pilling of fabric samples is taken firstly even if the methods are different. This can be done with the aid of a microscope, scanner, camera or an improved device. Next, image processing steps are performed in this image or video to determine the pills more easily. The characteristics of the pills, which are made more visible in the image, are determined. Afterwards, the pilling grades is tried to be determined with different statistical techniques by using pill characteristics.

The vast majority of previous studies were based on the existing standard photographs used for objective assessment of pilling. The new system to be developed in the studies in the literature was planned to be compatible with the existing systems. However, earlier image processing researches has shown that standard photographs of different organizations are incompatible even among themselves. It is often difficult to distinguish the pill and fabric structure in these images. Standard photographs are reproduced outputs of photographs of pilling of fabric samples under certain conditions such as different lighting, settlement, scale, contrast, etc. Good pill-background contrast in standard images is basically two types. The first is the light-dark contrast that separates the pills from the fabric background. This contrast is achieved when the sample is illuminated under horizontal light. In the second type of contrast, the fabric background appears as a high light-dark contrast, and the pills can be distinguished using regional changes in the form of pills (light) -shade (dark). This type of contrast is more difficult to achieve with accurate image analysis. The average light intensity is not constant in the photographs; some seem darker than others. Some images are in horizontal and vertical directions, while others are in the form of a diagonal. In the some photos, some shadows increase the area covered by the pills. Horizontal lighting causes irregular light intensity in photographs. The fabric surface is similar in all photos, but the fabric structure and orientation are not the same. In some photos, there are some faults such as thick threads, faulty threads, wrinkles on the fabric surface and small stains that can be confused with small pills [3, 10-21].

In the recent years, Telli (2019b) reported that they achieved success in EMPA SN 198525 "W3" standard photographs. Researcher created equations for the pill characteristics obtained from image processing steps of standard photographs with the help of a curve fitting technique. The obtained equations were applied to the original image of the fabric. In the quality ranges created for each fabric, it was determined what value the pilled fabric correspond to. The same results with the subjective results were obtained in the data of the total pill area in the images and the mean of the matrix elements that are its textural expression. Each fabric has its own unique structure. There are different colors, brightness, fuzz, pattern or surface structure. It was emphasized that the original image must be used to eliminate this. This situation has been shown as the reason for the failure of many successful image processing studies in the literature [21].

To eliminate the differences caused by the fabric structure mentioned above, one of the remarkable points in recent studies is that more professional techniques are used in image capturing [22-23]. Some studies have focused on the evaluation of fuzz and pills on the surface separately [24-26]. Few studies have focused on the textural properties of the image in the numerical expression of the pill characteristics [21, 27]. Different image processing approaches for objective evaluation of pilling in the literature are available from the study listed by Furferi et al. (2015) [28].

In addition to these academic ideas and discussions, PillGradeTM Automated 3D Pilling & Fuzz Grading System has been a commercial application [29]. In 2003, a large polyester manufacturer determined the need for an objective pilling measuring device within the scope of a project carried out to improve the pilling resistance of staple polyester fibers. As a result of collaboration with an American image technology development company called LineTech, the automatic pilling evaluation system was developed and PillGradeTM System was offered commercially by SDL-Atlas. PillGradeTM is a three-dimensional automatic imaging system that detects, measures and counts the pills formed on the fabric surface. The system mainly includes a software program, a camera, back and front light source, two mirrors, a feeding table and moving rollers. The PillGradeTM device does not take a photo, but a video recording. The mirrors placed at different angles provide the view of the fabric both from the horizontal and top surface at the same time. The first image processing step is to apply the threshold value to the image for the fabric surface line. Then, black and white threshold values are applied to detect the pills and fuzz separately.

Depending on the degree of fuzziness that may occur on the fabric surface, the pills can be hidden by the fuzz. Furthermore, the patterns or the structure of the fabric can prevent the pills formed on the surface from being seen easily. These situations can be misleading factors for evaluation. To prevent this, the PillGradeTM system is based on the measurement of the height of each pill and the measurement of fuzziness during the movement of the fabric around the rotating spindle by capturing its horizontal plane. Besides, pill heights are also taken into consideration since pilling evaluation is three-dimensional. Apart from the total number and area of pills in the system, the size of the pills is also important. The largest pills (>Ø3.9 mm) are multiplied by a factor of 5.0 and the smallest pills (<Ø0.3 mm) are multiplied by a factor of 0.6. The number of weighted pills obtained after the coefficients is used in the formula below.

$$\text{Grade} = 6 - (1 + \text{GradeMultiplier} * (\text{WeightedPills/inch2}))^{\text{POWER}} \qquad (1)$$

where GradeMultiplier has defaulted to 0.75 and POWER is defaulted 0.45.

However, it is interesting that the system also offers the possibility to adapt the pilling grade formula to suit the subjective assessment of the user in any product. The evaluation made by the system can be changed by changing the coefficients of five different variables that form the curve determining the degree of pilling. Five different variables are Grade Multiplier, Power, Weighted Pill Tare, Detect Threshold, Fuzz to Pill Thresh Mux. Kirtay and Kayseri (2014) studied the relationship between PillGradeTM and expert operators. The mean results obtained from the view of three experts were compared with PillGradeTM data in two different detect threshold levels. High and low the pill detection sensitivity were obtained changing the detect threshold variable. Pilling grades showed an increase with the decrease of pill sensitivity. The strong correlation values were determined between subjective assessments and PillGradeTM results. However, different results were seen from the mean results of experts in both low sensitivity and high sensitivity [30].

IMAGE PROCESSING STUDIES IN MATLAB SOFTWARE

When academic researches and commercial practices are examined, it is clear that the current subjective methods will remain valid for now. However, it seems that the image processing studies have reached the level to help in the hesitating results of subjective methods. It is thought that the determination of the pilling grades will become completely objective with comprehensive researches to be carried out in the following years.

As an example, MATLAB software offers the opportunity to perform image processing. Studies can be done by writing code or using the toolbox for beginners. You can create your system without the need to write a code in the toolbox and with the codes to be produced from there; it is possible to reach the result in a single move.

Steps in the Usage of Image Processing Toolbox

Firstly, MATLAB software is opened from the computer. Figure 4 shows the MATLAB 2018a software opening screen.

The image toolbox is opened by typing the "imtool" code in the MATLAB command window. The image to be processed is opened from the file section of the toolbox. Then, it is transferred to the workspace with the "export to workspace" button. Meanwhile, an appropriate "image variable name" is given. For this case study, 100% Cotton Single Jersey fabric having 200g/m2 weight was used. Pilling process was realized for 2000 and 5000 turns with Martindale Test Instrument according to EN ISO 12945-2. Original and post-pilling images of the fabric were taken as 900×900 pixels at the same size in RGB format by using a digital scanner. The fabric images are presented in Figure 5.

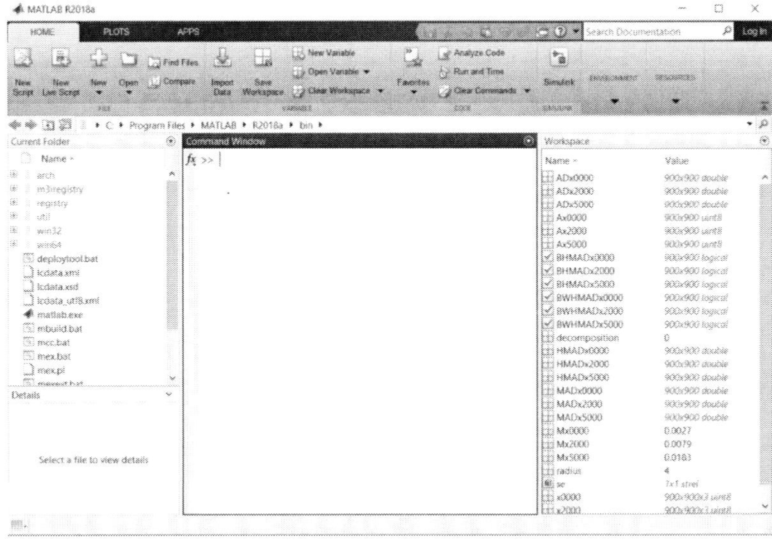

Figure 4. The MATLAB 2018a software opening screen.

Figure 5. Fabric images in RGB format.

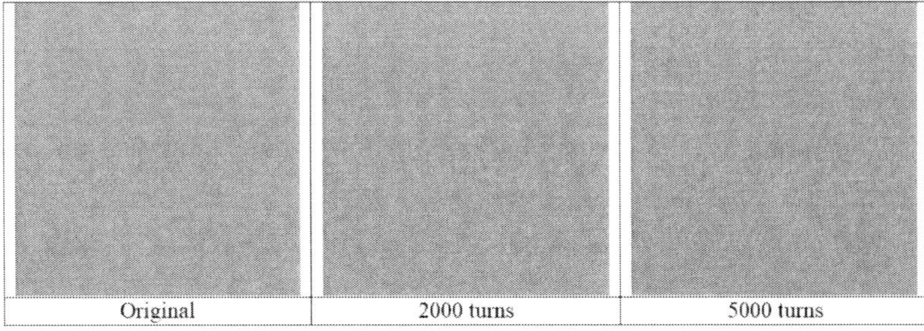

Figure 6. Fabric images in grayscale format.

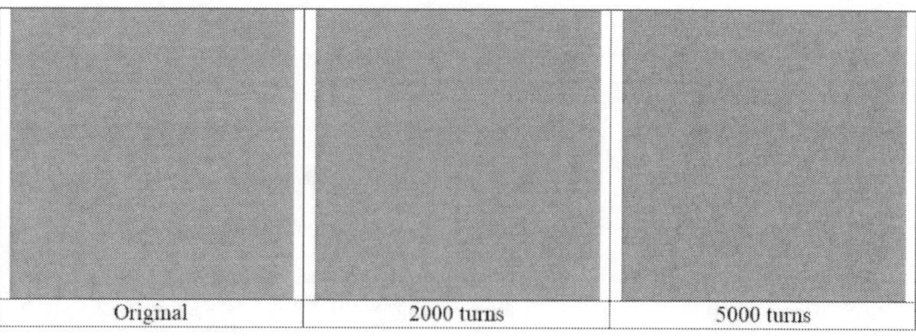

Figure 7. Fabric images in double format.

Then the codes of the desired image processing steps are written on the command line and the image is made ready for segmentation. For image segmentation, the three-dimensional matrix in RGB format must be converted to mxn double format. To do this, first of all, matrices are converted to grayscale format using"rgb2gray" code (Figure 6). Then the matrix is brought to the double format using the "im2double"code (Figure 7).

After this point, appropriate image processing techniques should be determined to reduce noise in the image and make the pills more distinct. In this study, the 2D median filtering technique recommended by Telli (2019c) for the detection of pilling was applied using the code "medfilt2" [31]. The obtained images after 2D median filtering are seen in Figure 8. To compare the three fabrics, after filtering, the histogram-fitting operation suggested by Xin (2002) was done using the "imadjust" code [15]. Figure 9 presents images after the histogram-fitting operation.

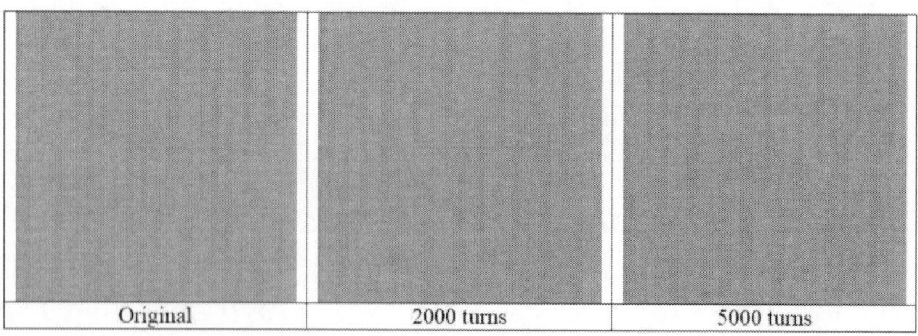

Figure 8. The obtained images after 2D median filtering.

Figure 9. Images after the histogram-fitting operation.

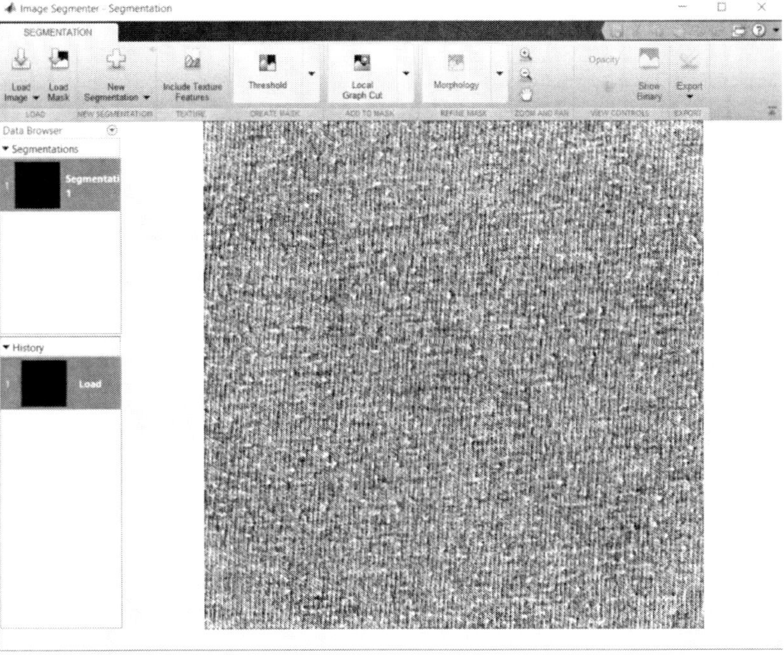

Figure 10. An image from "Image Segmenter" screen.

Then, these images were segmented using "Image Segmenter" in "Image Processing Toolbox". An image from "Image Segmenter" screen is given in Figure 10.

Segmented images by automatic global Otsu's algorithm are shown in Figure 11. After this stage, pills gained a clearer image compared to the background of the image.

Figure 11. Images after thresholding.

Then, values below the specified points are refined using differently shaped structuring element with the "morphology" button in the "Image Segmenter" toolbox. An image from "Morphology" screen is given in Figure 12.

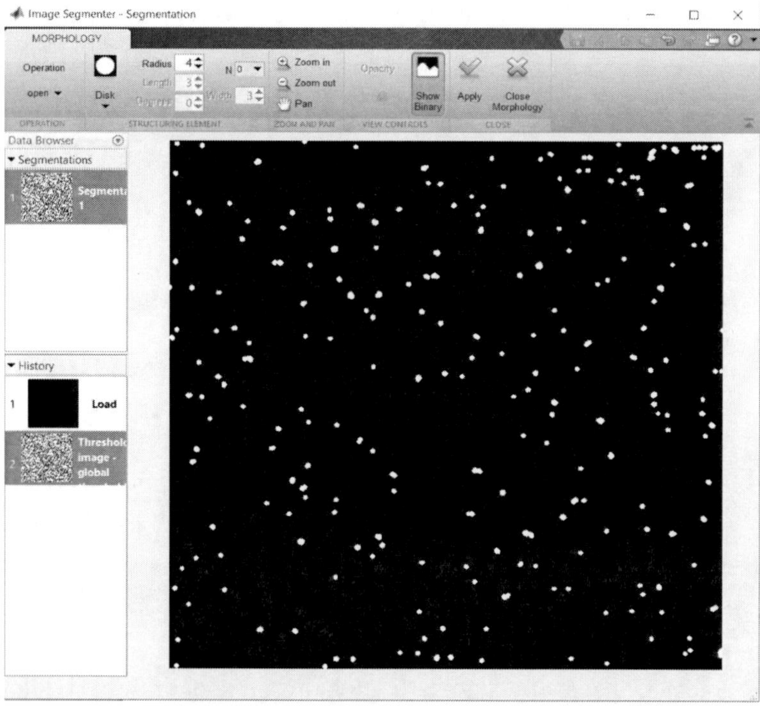

Figure 12. An image from "Morphology" screen in Image Segmenter.

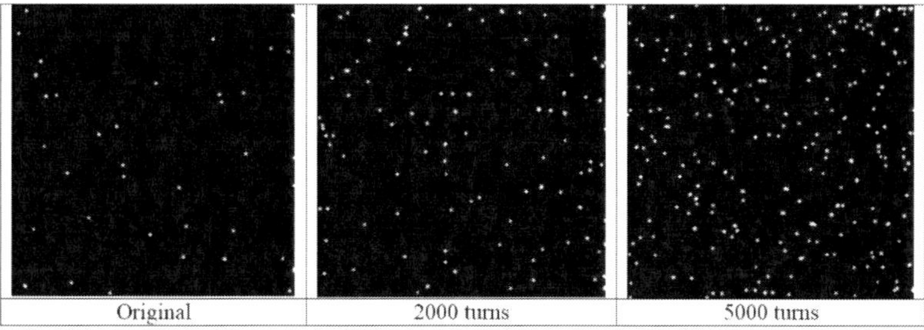

Figure 13. Images after morphological operations.

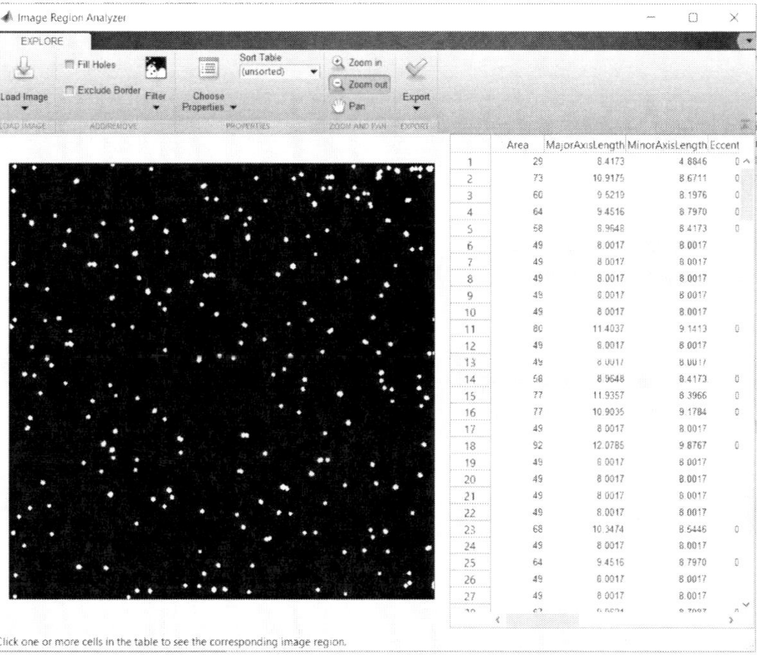

Figure 14. An image from “Image Region Analyzer” screen.

In morphological operations, four different masks (dilated, erode, open and close) can be selected. Also, the shape of the structuring element (disk, diamond, line, octagon, square or rectangle) can be changed according to the shape of the section to be segmented. And depending on the situation, their radius, length or degrees can be selected. In this study, “open-mask” was preferred due to the fabric structure. In an open mask, erosion followed by dilation is used. This removes foreground objects smaller than the structuring element. The shape of the structuring element has been selected as "disk" as we have detected pills. In this case, the radius value, which gives the minimum pill value in the original fabric, was taken as "4". Figure 13 shows images after morphological operations.

After this point, statistical properties can be extracted from the image matrix. This can be done in two ways. The first is to examine the properties of the pills in the image. The second is to examine the textural properties of the image directly. Pill properties can be examined in detail by using "Image Region Analyzer" in MATLAB software. Twelve different statistical values can be researched and filtered in detail. An image from "Image Region Analyzer" screen was given in Figure 14. To examine the textural features, the relevant code must be entered on the MATLAB command screen.

The vast majority of researchers focused on pill characteristics in image processing studies. In a limited number of studies, textural properties were examined [21, 27]. It has been stated in previous studies that it may be useful to use "mean of matrix elements" from textural properties and "total area" data from pill characteristics. In fact, these two parameters lead to the same point because "mean of matrix elements" is obtained by dividing the total area covered by the beads in the image by the entire pixel amount in the image. Therefore, in this study, "mean of matrix elements" values were obtained for each fabric using the "mean2" code. The results obtained in Table 3 are shown.

Without using Image Processing Toolbox, it is possible to reach the desired result in seconds by typing the code line as below in the command window.

```
% To convert images uploaded to MATLAB software to Grayscale format
>> Ax0000=rgb2gray (x0000);
>> Ax2000=rgb2gray (x2000);
>> Ax5000=rgb2gray (x5000);

% To convert Grayscale format to double format
>> ADx0000= im2double (Ax0000);
>> ADx2000= im2double (Ax2000);
```

```
>> ADx5000= im2double (Ax5000);

% Application of 2D Median Filtering
>> MADx0000= medfilt2 (ADx0000);
>> MADx2000= medfilt2 (ADx2000);
>> MADx5000= medfilt2 (ADx5000);

% Application of Histogram-Fitting
>> HMADx0000= imadjust (MADx0000);
>> HMADx2000= imadjust (MADx2000);
>> HMADx5000= imadjust (MADx5000);

% To start segmentation processing by thresholding
>> BHMADx0000 = imbinarize(HMADx0000);
>> BHMADx2000 = imbinarize(HMADx2000);
>> BHMADx5000 = imbinarize(HMADx5000);

% To finish segmentation processing by morphological operations
>> radius = 4; decomposition = 0; se = strel('disk', radius, decomposition);
BWHMADx0000 = imopen(BHMADx0000, se);
>> radius = 4; decomposition = 0; se = strel('disk', radius, decomposition);
BWHMADx2000 = imopen(BHMADx2000, se);
>> radius = 4; decomposition = 0;
se = strel('disk', radius, decomposition);
BWHMADx5000 = imopen(BHMADx5000, se);

% To compare by finding the mean of the matrix elements
>> Mx0000= mean2 (BWHMADx0000);
>> Mx2000= mean2 (BWHMADx2000);
>> Mx5000= mean2 (BWHMADx5000);
```

In addition, the results obtained from fabric samples evaluated subjectively by a specialist are given in Table 3.

Table 3. Mean of matrix elements results of images and the results of subjective evaluation

	Original	2000 turns	5000 turns
Mean of matrix elements	0,0027	0,0079	0,0183
The results of subjective evaluation	5	4	3

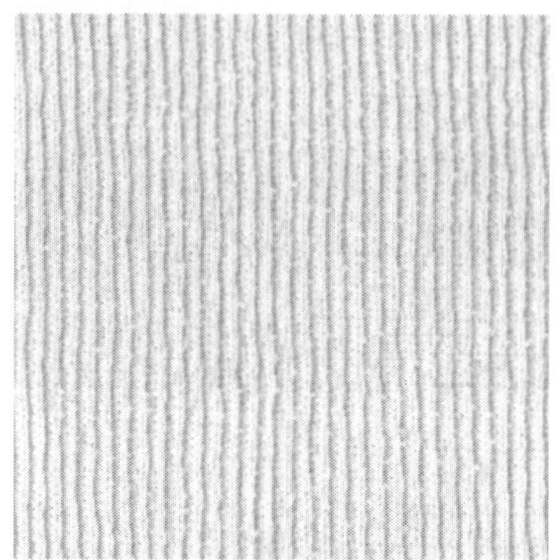

Figure 15. Rib knitted fabric having thick and wide wale loop.

Figure 16. Viscose single jersey fabrics produced from fancy yarns.

When Figure 5, Figure 13 and Table 3 are examined together, it is seen for this fabric that the number of pilling increases as the number of turns increases. It is observed that the values of the mean of matrix elements increase with the increasing amount of pilling. As mentioned in previous studies, it can be seen that there is a strong relationship between subjective evaluation results and "mean of matrix elements" values. However, all these are not sufficient for an objective pilling grading system. The different fabric structures make difficult the system. Many factors such as pattern structure, fabric surface structure, brightness, etc. can mislead the software. Figure 15 shows the rib knitted fabric structure that can mislead the software with its thick and wide wale loop.

Therefore, extra effort will be required in both image acquisition and processing. Wale loops are very thick and large such that the form of pills remains smaller than the wale. The pills are intertwined with the fabric background. For this reason, the segmentation of images and morphological operations become significantly difficult. The same situation is valid for viscose single jersey fabrics produced by using fancy yarns given in Figure 16.

A special effort is required to separate the pills from the image background in fabrics produced from mélange yarn seen on the left and fabrics produced from slub yarn seen on the right although their subjective pilling grades are "2". In this style, the preparation of image segmentation and morphological processes is an important problem.

However, the studies carried out up to now and the example study described above show that it is possible to make an objective pilling detection easily for the fabric structures used as standard in the textile industry thanks to the databases to be created with measuring lots of samples.

CONCLUSION

Pilling is a critical problem for fabrics. The pill formation causes an unpleasant surface and reduces the life of the fabric. Test results of pilling resistance play an important role in determining the quality and economic life of the garment. The determination of the pilling grades is made with visual control by operators. The human factor is significantly effective in this subjective evaluation method and may cause incorrect results. In general, it can be said that there is a strong correlation between experts in the determination of the pilling grades with conventional evaluation methods. However, differences in visual perception between experts make it difficult to obtain the same results. Similar findings were provided with the decrease of sensitivity with 0.5 units. However, differences in 0.5 units can be tolerated at high scores such as 5, 4/5 or 4. Serious problems may occur at the values below, depending on the desired quality value. Therefore, further works should be intensified on objective methods such as image processing in the evaluation of pilling.

Basically, in objective evaluation, the image or video recording of pilling of the fabric samples is taken firstly even if the methods are different. This can be done with the aid of a microscope, scanner, camera or an improved device. Next, image processing steps are performed in this image or video to determine the pills more easily. The characteristics of the pills, which are made more visible in the image, are determined. Afterwards, the pilling grades is tried to be determined with different statistical techniques by using pill characteristics. However, the different fabric structures make difficult the developed systems. Many factors such as pattern structure, fabric surface structure, brightness and etc. can mislead the software. In this context, the preparation of image segmentation and morphological processes is an important problem.

When academic researches and commercial practices are examined, it is clear that the current subjective methods will remain valid for now. However, it seems that the image processing studies have reached the level to help in the hesitating results of subjective methods especially easily for the fabric structures used as standard in the textile industry.

REFERENCES

[1] Ukponmwan, J. O., Mukhopadhyay, A., & Chatterjee, K. N. (1998). Pilling. *Textile progress*, 28(3), 1-57.

[2] Ozdil, N., (2003). Physical Quality Control Methods in Fabrics, *Ege University TEKAUM* Publication No: 21, ISBN No: 975-483-579-9, Bornova-İzmir, 120p.

[3] Ozcelik, G., (2009). A Research on the Objective Evaluation and Prediction of Fabric Pilling Property, *Ege University PhD Thesis*, Bornova-İzmir, 291p.

[4] Kayseri, G. Ö., & Kirtay, E. (2015). Part 1. Predicting the pilling tendency of the cotton interlock knitted fabrics by regression analysis. *Journal of Engineered Fibers and Fabrics*, *10*(3), 155892501501000305.

[5] Kayseri, G. Ö., & Kirtay, E. (2015). Part II. Predicting the pilling tendency of the cotton interlock knitted fabrics by artificial neural network. *Journal of Engineered Fibers and Fabrics*, *10*(4), 155892501501000417.

[6] Bozdogan, F., (2009). Physical Textile Control (Fabric Tests), *Ege University TEKAUM* Publication No: 32, Bornova-İzmir, 162p.

[7] Telli A., The Relationship between Yarn Properties and Pilling Resistance in Knitted Fabrics, *3rd International Symposium on Innovative Approaches in Scientific Studies (ISAS 2019)*, 19-21 Apr 2019, pp. 238-240.

[8] ISO 12945-2:2000 Textiles- Determination of fabric propensity to surface fuzzing and to pilling- Part 2: Modified Martindale method.

[9] Telli A., "The reliability of subjective assessment in the determination of pilling resistance of knitted fabrics," *4th International Mediterranean Science and Engineering Congress (IMSEC 2019)*, 25-27 Apr 2019, pp. 420-422

[10] Konda, A., Xin, L. C., Takadera, M., Okoshi, Y., & Toriumi, K. (1990). Evaluation of pilling by computer image analysis. *Journal of the textile Machinery Society of Japan*, *36*(3), 96-107.

[11] Xu, B. (1997). Instrumental evaluation of fabric pilling. *The Journal of The Textile Institute*, *88*(4), 488-500.

[12] Abril, H. C., Millan, M. S., Torres, Y., & Navarro, R. (1998). Automatic method based on image analysis for pilling evaluation in fabrics. *Optical Engineering*, *37*(11), 2937–2947.

[13] Hsi, C. H., Bresee, R. R., & Annis, P. A. (1998). Characterizing fabric pilling by using image-analysis techniques. Part I: Pill detection and description. *Journal of the Textile Institute*, *89*(1), 80-95.

[14] Hsi, C. H., Bresee, R. R., & Annis, P. A. (1998). Characterizing fabric pilling by using image-analysis techniques. Part II: Comparison with visual pill ratings. *Journal of the Textile Institute*, *89*(1), 96-105.

[15] Xin, B., Hu, J., & Yan, H. (2002). Objective evaluation of fabric pilling using image analysis techniques. *Textile Research Journal*, *72*(12), 1057-1064.

[16] Kang, T. J., Cho, D. H., & Kim, S. M. (2004). Objective evaluation of fabric pilling using stereovision. *Textile research journal*, *74*(11), 1013-1017.

[17] Palmer, S., & Wang, X. (2004). Evaluating the robustness of objective pilling classification with the two-dimensional discrete wavelet transform. *Textile Research Journal*, *74*(2), 140-145.

[18] Kim, S. C., & Kang, T. J. (2005). Image analysis of standard pilling photographs using wavelet reconstruction. *Textile Research Journal*, *75*(12), 801-811.

[19] Behera, B. K., & Mohan, T. M. (2005). Objective measurement of pilling by image processing technique. *International Journal of Clothing Science and Technology*. *17*(5), 279-291.

[20] Zhang, J., Wang, X., & Palmer, S. (2007). Objective pilling evaluation of wool fabrics. *Textile research journal*, *77*(12), 929-936.

[21] Telli, A. (2019). An Image Processing Research Consistent with Standard Photographs to Determine Pilling Grade of Woven Fabrics. *Tekstil ve Konfeksiyon*, *29*(3), 268-276.

[22] Techniková, L., Tunák, M., & Janáček, J. (2016). Pilling evaluation of patterned fabrics based on a gradient field method. *Indian Journal of Fibre & Textile Research (IJFTR)*, *41*(1), 97-101.

[23] Techniková, L., Tunák, M., & Janáček, J. (2017). New objective system of pilling evaluation for various types of fabrics. *The Journal of The Textile Institute*, *108*(1), 123-131.

[24] Cherkassky, A., & Weinberg, A. (2010). Objective evaluation of textile fabric appearance part 1: Basic principles, protrusion detection, and parameterization. *Textile Research Journal*, *80*(3), 226-235.

[25] Cherkassky, A., & Weinberg, A. (2010). Objective Evaluation of Textile Fabric Appearance. Part 2: SET Opti-grade Tester, Grading Algorithms, and Testing. *Textile Research Journal*, *80*(2), 135-144.

[26] Wang, L., Ouyang, W., Gao, W., & Xu, B. (2017). Instrumental evaluation of fabric abrasive wear using 3D surface images. *The Journal of The Textile Institute*, *108*(5), 846-851.

[27] Eldessouki, M., & Hassan, M. (2015). Adaptive neuro-fuzzy system for quantitative evaluation of woven fabrics' pilling resistance. *Expert Systems with Applications*, *42*(4), 2098-2113.

[28] Furferi, R., Governi, L., & Volpe, Y. (2015). Machine vision-based pilling assessment: a review. *Journal of Engineered Fibers and Fabrics*, *10*(3), 155892501501000320.

[29] SDL Atlas, PillGrade, https://sdlatlas.com/products/pillgrade-automated-pilling-grading-system (05.07.2020).

[30] Kirtay, E., & Kayseri, G. Ö. (2014). Subjective and objective evaluation of pilling. *Engineering (ICITE 2014)*, *1*, 39.

[31] Telli A., The Usage of Different Filter in Processing of Pilling Images. *The International Conference of Materials and Engineering Technologies (TICMET'19)*, 10-12 October 2019, pp. 695-701.

Index

A

B

C

D

E

F

G

H

I

P

Q

R

S

T

U

V

W

X

Y

Z